MICROBIOLOGICALLY SAFE FOODS

ABOUT THE AUTHOR

Dr. Linex Wolper, M.Sc. Ph.D. in Microbiology is Asst. Professor on Adhoc basis at RWTH Aachen University-Germany (Templergraben). Linex is teaching in this college for last four years. He is also a visiting faculty at reputed IAS Academy in Germany (Aachen). He has attended three International level seminars In his carrier. He was teaching in University of Roehampton in London for 2 years. Linex has published his two articles also with International Publisher. Linex is a man with a passion for rural development especially in the area of eradication of rural poverty, using dairy as an accelerator.

ABOUT THE BOOK

Food microbiology is the study of the microorganisms that inhabit, create, or contaminate food. Bacteria are incredibly diverse and abundantly found in most of the natural world. The majority are beneficial to us in ways we may not fully realize or appreciate. A few, however, are not and will cause disease when we cross paths with them. Pathogenic (harm-causing) and potentially pathogenic bacteria may be found in unexpected places, such as in the food we eat, the water we drink or use for recreation, in soil, on surfaces in your home, and elsewhere. Microbiology is the science which includes the study of the occurrence and significance of bacteria, fungi, protozoa and algae which are the beginning and ending of intricate food chains upon which all life depends. These food chains begin wherever photosynthetic organisms can trap light energy and use it to synthesize large molecules from carbon dioxide, water and mineral salts forming the proteins, fats and carbohydrates which all other living creatures use for food. Within and on the bodies of all living creatures, as well as in soil and water, micro-organisms build up and change molecules, extracting energy and growth substances. This book focuses on state of the art technologies to produce microbiologically safe foods for our global dinner table.

MICROBIOLOGICALLY SAFE FOODS

Dr. Linex Wolper

WESTBURY PUBLISHING LTD.
ENGLAND (UNITED KINGDOM)

Microbiologically Safe Foods
Edited by Dr. Linex Wolper
ISBN: 978-1-913229-57-3 (Hardback)

© 2020 Westbury Publishing Ltd.

Published by **Westbury Publishing Ltd.**
Address: 6-7, St. John Street, Mansfield,
Nottinghamshire, England, NG18 1QH
United Kingdom
Email: - info@westburypublishing.com
Website: - www.westburypublishing.com

This book contains information obtained from authentic and highly regarded sources. All chapters are published with permission under the Creative Commons Attribution Share Alike License or equivalent. A Wide Variety of references are listed. Permissions and sources are indicated; for detailed attributions, please refer to the permission page. Reasonable efforts have been made to publish reliable data and information, but the authors, editors and publisher cannot assume any responsibility for the validity of the materials or the consequences of their use.

The publisher's policy is to use permanent paper from mills that operate a sustainable forestry policy. Furthermore, the publishers ensure that the text paper and cover boards used have met acceptable environmental accreditation standards.

Publisher Notice: - Presentations, Logos (the way they are written/ Presented), in this book are under the copyright of the publisher and hence, if copied/ resembled the copier will be prosecuted under the law.

British Library Cataloguing in Publication Data:
A catalogue record for this book is available from the British Library.

For more information regarding Westbury Publishing Ltd and its products, please visit the publisher's website- **www.westburypublishing.com**

Preface

Food microbiology is the study of the microorganisms that inhabit, create, or contaminate food.

Unfortunately for us, the things we eat and drink are fairly common vehicles for disease transmission. And, because food and drink pass through our digestive tract, the most common symptoms of a foodborne disease are abdominal discomfort or pain, nausea, diarrhea, and/or vomiting. Gastrointestinal illnesses caused by foodborne microbes range in severity from mild to extremely debilitating, even fatal. The biological agents responsible for this type of disease may be viruses, bacteria, fungi, protozoa, or helminthes.

The focus of Food Microbiology is on the detection and analysis of foodborne spoilage microorganisms. Food microbiology is the study of food micro-organisms; how we can identify and culture them, how they live, how some infect and cause disease and how we can make use of their activities. Microbes are single-cell organisms so tiny that millions can fit into the eye of a needle. They are the oldest form of life on earth. Microbe fossils date back more than 3.5 billion years to a time when the Earth was covered with oceans that regularly reached the boiling point, hundreds of millions of years before dinosaurs roamed the earth.

The field of food microbiology is a very broad one, encompassing the study of microorganisms which have both beneficial and deleterious effects on the quality and safety of raw and processed foods. Food science is a discipline concerned with all aspects of food-beginning after harvesting, and ending with consumption by the consumer. It is considered one of the agricultural sciences, and it is a field which is entirely distinct from the field of nutrition. In the U.S., food science is typically studied at land-grant universities. Examples of the activities of food scientists include the development of new food products, design of processes to produce these foods, choice of packaging materials, shelf-life studies, sensory evaluation of the product with potential consumers, microbiological and chemical testing. Food scientists in universities may study more fundamental phenomena that are directly linked to the production of a particular food product. Food scientists are generally not directly involved with the creation of genetically modified (bio-engineered) foods. Some of the subdisciplines of food science: Food safety, Food engineering, Product development, Sensory analysis, Food chemistry.

The primary tool of microbiologists is the ability to identify and quantitate food-borne microorganisms; however, the inherent inaccuracies in enumeration processess, and the natural variation found in all bacterial populations complicate the microbiologists job. Without microbes, we couldn't eat or breathe. Without us, they'd probably be just fine.

Microorganisms are small, living organisms including yeasts, molds, bacteria, and parasites. Some are beneficial and allow for the production of bread, cheese, wine and antibiotics. Others cause foods to spoil (e.g., mould) and yet others, called pathogens, make people sick by producing toxins or poisons. Microbes are found everywhere and can be moved to food by insects, utensils, equipment, hands, air, dust, and water.

This is a reference book. All the matter is just compiled and edited in nature, taken from the various sources which are in public domain.

This book focuses on state of the art technologies to produce microbiologically safe foods for our global dinner table.

—*Editor*

Contents

Preface (v)

1. **Introduction** 1
   - Hepatitis A ; Natural Toxins; Mechanism; Society and culture; Changes in foods caused by Chryseobacterium ; Growth And Hydrolytic Activities Of Chryseobacterium Species In Milk ; Food Borne Bacteria; Gram Positive Rods And Cocci; C. botulinum

2. **Foods from Microorganisms** 23
   - Dairy foods ; Other fermented foods ; Why Do We Need to Preserve Foods; Food Preservation with Home Canning Halts Spoilage; Food Safety From Microorganisms ; Microbial deterioration of food components

3. **Food Safety** 39
   - Introduction; Thermal Processing of Foods; Hazard Classification ; Contamination Levels and Microbiological Control; Food Safety Considerations; Safety Assessment; Special Issues; Voluntary Food Safety Management System

4. **Food Spoilage** 84
   - Characteristics Of Food Spoilage; Types of spoilage ; Microbiological Aspects Of Foods ; Microbial metabolites; Growth of food spoilage bacteria; The genus Chryseobacterium; Spoilage caused by flavobacteria

5. **Scope of Food Microbiology** 110
   - Basic Of Microbiology ; The history of microbiology; Human effected by microbes; Benefits of Microbes; Environment effected by microbes; Microbes around us; Spectrum of Microbiology; Industrial microbiology/biotechnology?; Microorganisms And Food ; Growth Of Microorganisms

6. **Prevention of Spoilage in Milk**   133
   - Drink of Milk Product Items; Hygiene during Milking; Pasteurised milk products; Pathogens: Growth and Survival; Factors Affecting Spoilage; Micro-organisms Action

7. **Food Applications for Human Health**   165
   - Quantity and Quality of Food Oils; Biotechnology and Global Health; Risk Of GMOs And GM Foods To Human Health And The Environment

8. **Chemical Hazards Associated with the Food Supply**   198
   - Cause for Increasing Concern

9. **Foodborne Pathogens**   208
   - Common Foodborne Pathogens; Foodborne Disease Outbreaks; Food Preservation ; Impact of Intellectual Property Rights on Research

10. **Defining Food Security**   235
    - The Challenges to Food Security; Climate Change and Food Security; Quantifying the Impacts on Food Security; Risks to food security; Essentially of Food Security; Ensuring Global Food Security

11. **Meat and Poultry Hazard Analysis**   265

    Bibliography   269

    Index   273

1
Introduction

Health experts estimate that the yearly cost of all foodborne diseases in this country is 5 to 6 billion dollars in direct medical expenses and lost productivity. There are more than 250 known foodborne diseases. They can be caused by bacteria, viruses, or parasites. Natural and manufactured chemicals in food products also can make people sick. Some diseases are caused by toxins from the disease-causing microbe, others by the human body's reactions to the microbe itself.

To better understand the epidemiology of foodborne diseases in the United States, 10 states across the country are collecting annual data on the occurrence of new cases of the most common causes of bacterial and parasitic infections through the Foodborne Diseases Active Surveillance Network, a CDC-sponsored programme known as FoodNet. Recently, public health, agriculture, and environmental officials have expressed growing concern about keeping the nation's food and water supply safe from terrorist acts of introducing foodborne microbes. A number of US agencies, including the National Institutes of Health, CDC, Food and Drug Administration, US Department of Agriculture, and US Environmental Protection Agency, are studying this bioterrorism threat.

HEPATITIS A

Hepatitis A is an acute infectious disease of the liver caused by the hepatitis A virus which is transmitted person-to-person by ingestion of contaminated food or water or through direct contact with an infectious person. Tens of millions of individuals worldwide are estimated to become infected with HAV each year. The time between infection and the appearance of the symptoms is between two and six weeks and the average incubation period is 28 days.

In developing countries, and in regions with poor hygiene standards, the incidence of infection with this virus is high and the illness is usually contracted in early childhood. As incomes rise and access to clean water increases, the incidence of HAV decreases. Hepatitis A infection causes no clinical signs and symptoms in over 90 per cent of infected children and since the infection confers lifelong immunity, the disease is of no special significance to those infected early in life. In Europe, the United States and other industrialized countries, on the other hand, the infection is contracted primarily by susceptible young adults, most of whom are infected with the virus during trips to countries with a high incidence of the disease or through contact with infectious persons.

HAV infection produces a self-limited disease that does not result in chronic infection or chronic liver disease. However, 10 per cent-15 per cent of patients might experience a relapse of symptoms during the 6 months after acute illness. Acute liver failure from Hepatitis A is rare. The risk for symptomatic infection is directly related to age, with >80 per cent of adults having symptoms compatible with acute viral hepatitis and the majority of children having either asymptomatic or unrecognized infection. Antibody produced in response to HAV infection persists for life and confers protection against reinfection. The disease can be prevented by vaccination, and hepatitis A vaccine has been proven effective in controlling outbreaks worldwide.

Signs and symptoms

Early symptoms of hepatitis A infection can be mistaken for influenza, but some sufferers, especially children, exhibit no symptoms at all. Symptoms typically appear 2 to 6 weeks, after the initial infection.

Symptoms can return over the following 2-6 months and include:
- Fatigue,
- Fever,
- Abdominal pain,
- Nausea,
- Diarrhoea,
- Appetite loss,
- Depression,
- Jaundice, a yellowing of the skin or whites of the eyes,
- Sharp pains in the right-upper quadrant of the abdomen,

Introduction

- Weight loss,
- Itching,
- Bile is removed from blood stream and excreted in urine, giving it a dark amber colour, and
- Feces tend to be light in colour due to lack of bilirubin in bile.

Pathogenesis

Following ingestion, HAV enters the bloodstream through the epithelium of the oropharynx or intestine. The blood carries the virus to its target, the liver, where it multiplies within hepatocytes and Kupffer cells. Virions are secreted into the bile and released in stool. HAV is excreted in large quantities approximately 11 days prior to appearance of symptoms or anti-HAV IgM antibodies in the blood. The incubation period is 15-50 days and mortality is less than 0.5 per cent. Within the liver hepatocytes the RNA genome is released from the protein coat and is translated by the cell's own ribosomes.

Unlike other members of the Picornaviruses this virus requires an intact eukaryote initiating factor 4G for the initiation of translation. The requirement for this factor results in an inability to shut down host protein synthesis unlike other picornaviruses. The virus must then inefficiently compete for the cellular translational machinery which may explain its poor growth in cell culture. Presumably for this reason the virus has strategically adopted a naturally highly deoptimized codon usage with respect to that of its cellular host. Precisely how this stategy works is not quite clear yet. There is no apparent virus-mediated cytotoxicity presumably because of the virus' own requirement for an intact eIF4G and liver pathology is likely immune-mediated.

Diagnosis

Although HAV is excreted in the feces towards the end of the incubation period, specific diagnosis is made by the detection of HAV-specific IgM antibodies in the blood. IgM antibody is only present in the blood following an acute hepatitis A infection. It is detectable from one to two weeks after the initial infection and persists for up to 14 weeks. The presence of IgG antibody in the blood means that the acute stage of the illness is past and the person is immune to further infection. IgG antibody to HAV is also found in the blood following vaccination and tests for immunity to the virus are based on the detection of this antibody. During the acute stage of the infection, the liver enzyme alanine transferase is present in the blood at levels much higher than is normal. The enzyme comes from the liver cells that have been damaged by the virus. Hepatitis A virus is present in the blood, and feces of infected people up to two weeks before clinical illness develops.

Prevention

Hepatitis A can be prevented by vaccination, good hygiene and sanitation. The vaccine protects against HAV in more than 95 per cent of cases for longer than 20 years. It contains inactivated hepatitis A virus providing active immunity against a future infection. The vaccine was first phased in 1996 for children in high-risk areas, and in 1999 it was spread to areas with elevating levels of infection. The vaccine is given by injection. An initial dose provides protection starting two to four weeks after vaccination; the second booster dose, given six to twelve months later, provides protection for over twenty years.

Treatment

There is no specific treatment for hepatitis A. Sufferers are advised to rest, avoid fatty foods and alcohol, eat a well-balanced diet, and stay hydrated. Pharmacotherapeutic goals are to reduce the morbidity and prevent comlications involved in infection.

The therapy is given by the agents like:
- Analgesics—to reduce the abdominal pain; usually Acetaminophen is given. Acetaminophen also acts as an antipyretic, to reduce the fever.
- Antiemetics—to supress vomiting and nausea; usually Metoclopramide is given.
- Immune globulins usually BayGam 15-18 per cent IG is given through intramuscular route.

The United States Centres for Disease Control and Prevention in 1991 reported a low mortality rate for hepatitis A of 4 deaths per 1000 cases for the general population but a higher rate of 17.5 per 1000 in those aged 50 and over.

The risk of death from acute liver failure following HAV infection increases with age and when the person has underlying chronic liver disease. Young children who are infected with hepatitis A typically have a milder form of the disease, usually lasting from 1–3 weeks, whereas adults tend to experience a much more severe form of the disease.

Epidemiology
Prevalence

Antibodies to HAV in the blood are a marker of past or current infection. High-income regions have very low HAV endemicity levels and a high proportion of susceptible adults, low-income regions have high endemicity levels and almost no susceptible adolescents and adults, and most middle-income regions have a mix of intermediate and low endemicity levels. Anti-HAV prevalence suggest that middle-income regions in Asia, Latin America, Eastern Europe, and the Middle East currently have an intermediate or low level of endemicity. The countries in these regions may have an increasing burden of disease from hepatitis A. There were 30,000 cases of Hepatitis A reported to the CDC in the US in 1997. The agency estimates that there were as many as 2,70,000 cases each year from 1980 through 2000.

Transmission

The virus spreads by the fecal-oral route and infections often occur in conditions of poor sanitation and overcrowding. Hepatitis A can be transmitted by the parenteral route but very rarely by blood and blood products. Foodborne outbreaks are not uncommon, and ingestion of shellfish cultivated in polluted water is associated with a high risk of infection. Approximately 40 per cent of all acute viral hepatitis is caused by HAV. Infected individuals are infectious prior to onset of symptoms, roughly 10 days following infection. The virus is resistant to detergent, acid, solvents, drying, and temperatures up to 60°C. It can survive for months in fresh and salt water. Common-source outbreaks are typical. Infection is common in children in developing countries, reaching 100 per cent incidence, but following infection there is life-long immunity. HAV can be inactivated by: chlorine treatment, formalin, peracetic acid, beta-propiolactone, and UV radiation.

Cases

The most widespread hepatitis A outbreak in the United States afflicted at least 640 people in north-eastern Ohio and south-western Pennsylvania in late 2003. The outbreak was blamed on tainted green onions at a restaurant in Monaca, Pennsylvania. In 1988, 300,000 people in Shanghai, China were infected with HAV after eating clams from a contaminated river.

Virology

The Hepatitis virus is a Picornavirus; it is non-enveloped and contains a single-stranded RNA packaged in a protein shell. There is only one serotype of the virus, but multiple genotypes exist. Codon use within the genome is biased and unusually distinct from its host. It also has a poor internal ribosome entry site In the region that codes for the HAV capsid there are highly conserved clusters of rare codons that restrict antigenic variability.

NATURAL TOXINS

Several foods can naturally contain toxins, many of which are not produced by bacteria. Plants in particular may be toxic; animals which are naturally poisonous to eat are rare.

In evolutionary terms, animals can escape being eaten by fleeing; plants can use only passive defences such as poisons and distasteful substances, for example capsaicin in chili peppers and pungent sulfur compounds in garlic and onions. Most animal poisons are not synthesised by the animal, but acquired by eating poisonous plants to which the animal is immune, or by bacterial action.

- Alkaloids
- Ciguatera poisoning
- Grayanotoxin (honey intoxication)
- Mushroom toxins
- Phytohaemagglutinin (red kidney bean poisoning; destroyed by boiling)
- Pyrrolizidine alkaloids
- Shellfish toxin, including paralytic shellfish poisoning, diarrhetic shellfish poisoning, neurotoxic shellfish poisoning, amnesic shellfish poisoning and ciguatera fish poisoning
- Scombrotoxin
- Tetrodotoxin (fugu fish poisoning)

Some plants contain substances which are toxic in large doses, but have therapeutic properties in appropriate dosages.

- Foxglove contains cardiac glycosides.
- Poisonous hemlock (conium) has medicinal uses.
- Prions, resulting in Creutzfeldt-Jakob disease

Introduction

Ptomaine Poisoning

An early theory on the causes of food poisoning involved ptomaines ("fall, fallen body, corpse"), alkaloids found in decaying animal and vegetable matter. While some alkaloids do cause poisoning, the discovery of bacteria left the ptomaine theory obsolete, though as recently as 1882 the ptomaine was thought of as bacteria, while cadaverine and putrescine "special alkaloids" produced by the "comma bacillus".

MECHANISM

Incubation Period

The delay between consumption of a contaminated food and appearance of the first symptoms of illness is called the incubation period. This ranges from hours to days, depending on the agent, and on how much was consumed. If symptoms occur within 1–6 hours after eating the food, it suggests that it is caused by a bacterial toxin or a chemical rather than live bacteria. The long incubation period of many food borne illnesses tends to cause sufferers to attribute their symptoms to stomach flu. During the incubation period, microbes pass through the stomach into the intestine, attach to the cells lining the intestinal walls, and begin to multiply there. Some types of microbes stay in the intestine, some produce a toxin that is absorbed into the bloodstream, and some can directly invade the deeper body tissues. The symptoms produced depend on the type of microbe.

Infectious Dose

The infectious dose is the amount of agent that must be consumed to give rise to symptoms of food borne illness, and varies according to the agent and the consumer's age and overall health. In the case of Salmonella a relatively large inoculum of 1 million to 1 billion organisms is necessary to produce symptoms in healthy human volunteers, as Salmonellae are very sensitive to acid. An unusually high stomach pH level (low acidity) greatly reduces the number of bacteria required to cause symptoms by a factor of between 10 and 100.

SOCIETY AND CULTURE

Global Impact

Many outbreaks of food borne diseases that were once contained within a small community may now take place on global dimensions. Food safety authorities all over the world have acknowledged that ensuring food safety must not only be tackled at the national level but also through closer linkages among food safety authorities at the international level. This is important for exchanging routine information on food safety issues and to have rapid access to information in case of food safety emergencies. It is difficult to estimate the global incidence of food borne disease, but it has been reported that in the year 2000 about 2.1 million people died from diarrhoeal diseases. Many of these cases have been attributed to contamination of food and drinking water. Additionally, diarrhea is a major cause of malnutrition in infants and young children.

Even in industrialized countries, up to 30% of the population of people have been reported to suffer from food borne diseases every year. In the U.S, around 76 million cases of food borne diseases, which resulted in 325,000 hospitalizations and 5,000 deaths, are estimated to occur each year.

Developing countries in particular are worst affected by food borne illnesses due to the presence of a wide range of diseases, including those caused by parasites. Food borne illnesses can and did inflict serious and extensive harm on society.

In 1994, an outbreak of salmonellosis due to contaminated ice cream occurred in the USA, affecting an estimated 224,000 persons. In 1988, an outbreak of hepatitis A, resulting from the consumption of contaminated clams, affected some 300,000 individuals in China.

Food contamination creates an enormous social and economic strain on societies. In the U.S., diseases caused by the major pathogens alone are estimated to cost up to US $35 billion annually in medical costs and lost productivity.

The re-emergence of cholera in Peru in 1991 resulted in the loss of US $500 million in fish and fishery product exports that year. Food borne illness may carry long-term effects.

United Kingdom

In postwar Aberdeen a large scale outbreak of Typhoid occurred, this was caused by contaminated corned beef which had been imported from Argentina. The corned beef was placed in cans and because the cooling plant had failed, cold river water from the Plate estuary was used to cool the cans. One of the cans had a defect and the meat inside was contaminated. This meat was then sliced using a meat slicer in a shop in Aberdeen, and a lack of cleaning the machinery led to spreading the

Introduction

contamination to other meats cut in the slicer. These meats were then eaten by the people of Aberdeen who then became ill.

In the UK serious outbreaks of food-borne illness since the 1970s prompted key changes in UK food safety law. These included the death of 19 patients in the Stanley Royd Hospital outbreak and the bovine spongiform encephalopathy outbreak identified in the 1980s. The death of 17 people in the 1996 Wishaw outbreak of E. coli O157 was a precursor to the establishment of the Food Standards Agency which open and dedicated to the interests of consumers".

United States

In 1999 an estimated 5,000 deaths, 325,000 hospitalizations and 76 million illnesses were food borne in the US. In 2001, the Center for Science in the Public Interest petitioned the United States Department of Agriculture to require meat packers to remove spinal cords before processing cattle carcasses for human consumption, a measure designed to lessen the risk of infection by variant Creutzfeldt-Jakob disease.

The petition was supported by the American Public Health Association, the Consumer Federation of America, the Government Accountability Project, the National Consumers League, and Safe Tables Our Priority. This was opposed by the National Cattlemen's Beef Association, the National Renderers Association, the National Meat Association, the Pork Producers Council, sheep raisers, milk producers, the Turkey Federation, and eight other organizations from the animal-derived food industry. This was part of a larger controversy regarding the United States' violation of World Health Organization proscriptions to lessen the risk of infection by variant Creutzfeldt-Jakob disease. None of the US Department of Health and Human Services targets regarding incidence of food borne infections were reached in 2007.

CHANGES IN FOODS CAUSED BY CHRYSEOBACTERIUM

The non-diffusible, non-fluorescent bright yellow to orange pigments produced by most Chryseobacterium strains belong to the flexirubin type. The production of pigment may depend on the culture medium; it may also be more pronounced at low temperatures, in the presence of daylight, and in the presence of compounds such as casein, milk and starch. The production of extracellular slimy substances after prolonged incubation has been reported in *C. defluvii* and contributes to the increased hydration capacity that accompanies low-temperature meat spoilage. A strong odour, reminiscent of that emitted by cultures of some Empedobacter, Myroides and Sphingobacterium strains, is produced by most Chryseobacterium strains in liquid and solid culture; it has been described as "fruity".

Fruity odours arise from the degradation of the amino acids glycine, leucine and serine to form lower fatty acid esters. In a study by Engelbrecht *et al.* to determine spoilage characteristics of bacteria isolated from Cape hake, the odours produced were arranged into four offodour categories: stale, pungent, fruity and sulphidy. Thirty per cent of the Flavobacterium isolates produced no odours. Twenty-six per cent of the Flavobacterium isolates were able to produce H2S. Of the Flavobacterium balustinum strains tested for odour production, 75 per cent produced odours. Of the unidentified Flavobacterium-like isolates tested, 58 per cent produced odours.

The genus Flavobacterium had spoilage potential on the basis of off-odour production under the specified test conditions. Various proteolytic activities and H_2S production were also found in *Chryseobacterium balustinum*, *C. gleum* and *C. indologenes* strains isolated from Cape marine fish in South Africa. When offodour production of the Chryseobacterium strains was evaluated in fish muscle extract, four of the eight *C. balustinum* strains were found to produce a pungent odour, two a stale odour, and two no odour. Four of the five *C. gleum* strains produced a stale odour, and one *C. indologenes* strain produced a fruity odour.

Flavobacteria are generally considered to be of low significance in fish and other food spoilage. Off-odours such as sweet and fruity, putrid, sulphur and cheesy, characterize aerobically stored meat. Off-odours of poultry have been listed as sulphide-like, fruity, fishy, and like evapourated milk. Milk spoilage is associated with a bitter flavour and rancidity as well as a fruity odour. The fruity odour is due to a mixture of ethyl butyrate and ethyl hexanoate. Homospermidine is the major polyamine in *C. balustinum*, *C. gleum*, *C. meningosepticum* and *C. indologenes*, trace amounts of putrescine and agmatine are also detected in the three former species, whereas diaminopropane and 2-hydroxyputrescine are minor components in *C. indologenes*. Recently described Chryseobacterium species have not been investigated for their polyamine composition, except *C. defluvii*, which contains symhomospermidine as the major component and spermine and spermidine as minor components.

Introduction

GROWTH AND HYDROLYTIC ACTIVITIES OF CHRYSEOBACTERIUM SPECIES IN MILK

Several Chryseobacterium species which include *C. balustinum*, *C. gleum* and *C. joostei* are associated with the spoilage of dairy products during cold storage. In milk they produce heat resistant proteolytic and lipolytic enzymes responsible for off-flavours. They also produce off-odours in pasteurised milk and cream, surface taint in butter, thinning in creamed rice and bitterness due to the production of phospholipase C. The production of these spoilage enzymes are dependent *inter alia* on temperature and growth phase. Consumers depend upon their senses of sight, smell, taste and touch to evaluate food quality.

A careful combination of microbiological-, sensory- and chemical analysis are required to determine the spoilage potential of a microorganism. Despite the importance of microorganisms in food spoilage, the definition and assessment of spoilage relies on sensory evaluation since neither the "total count" of e.g., 107 cfu/ cm^2 nor the level of a specific spoilage organism *per se* can directly predict the sensory quality of a product. The growth and activity of spoilage microorganisms is mostly described and studied as a function of substrate base and of chemical and physical parametres such as temperature, pH, water activity and atmosphere.

The production of enzymes will always give valuable information on potential spoilage abilities while production of metabolites containing alcoholic compounds contribute to certain flavour compounds. Volatile and semi-volatile organic compounds present both in the matrix and the headspace aroma are largely responsible for the flavour qualities of the foods we eat. Flavour is considered to be the most important factor of any of the quality categories when comparing various dairy products. Dairy products provide a great sense of eating pleasure as a result of their characteristic flavour and smooth texture and they play an important role in a well-balanced diet. Because milk possesses a bland and soft flavour, the appearance of an objectionable off-flavour or off-odour is readily noticeable.

There is a delicate balance between many flavour compounds that contribute to a desirable milk flavour and if this balance is disturbed, off-flavours may occur. Flavobacteria can produce a putrid butter effect and a "sweaty feet" odour in skim milk. The aim of this study was to investigate the different growth and hydrolytic characteristics of the type species of the Chryseobacterium genus. These characteristics included the optimum growth temperature and growth capacity at different temperatures, pH values and NaCl concentrations. The proteolytic, lipolytic and phospholipase C production of the species was examined by using different microbiological culture media. The volatile compounds produced by the reference strains in milk, were examined by headspace gas chromatography. An odour description of milk samples, inoculated with the bacterial cultures were given by a semi-trained sensory panel of ten persons.

Materials and methods

Strains Investigated

Freeze-dried or agar slant cultures were reactivated in 10 ml Nutrient Broth and purity was checked by streaking on Nutrient Agar. Incubation was at 25°C for 24-48 h. The strains were maintained on Nutrient Agar slants and stored at 4°C.

Growth Characteristics

A battery of tests was performed to determine the different growth characteristics and to differentiate between Gram negative, yellow pigmented species, just as to the methods described by MacFaddin, Gerhardt *et al.*, Barrow and Feltham and Hugo *et al.*, unless indicated otherwise. All the tests were performed at 25°C, except where the growth of the reference strains at different temperatures, was examined on Nutrient Agar at 4°C, 25°C, 32°C, 37°C, 42°C and 55°C respectively.

Optimum Growth Temperature

The optimum growth temperature for four Chryseobacterium species was determined just as to the method described by Jooste *et al.* Four consecutive overnight cultures were prepared in tubes containing 10 ml each of Nutrient Broth. The Chryseobacterium species were incubated in a water-bath at each of the following temperatures: 20, 25, 30, 35, 40 and 45°C. Turbidity in the tubes, after 48 h, was read in a Milton Roy Spectronic 20D spectrophotometre at 660 nm using uninoculated Nutrient Broth as control. Optical density values were converted to Klett units.

Hydrolytic Activities

The proteolytic activity of the reference strains was determined by casein and gelatin hydrolysis, described by Harrigan and McCance. The

Introduction

lipolytic activity was determined using lecithinase production and the hydrolysis of olive oil and tributyrin. Phospholipase C production was determined using the method of Chrisope et al.

Preliminary Determination of Volatile Compounds in Milk

Growth conditions were as follows: 100 ml samples of Clover, UHT process, fat free milk and full cream milk were measured aseptically into sterile Erlenmeyer flasks. Each milk sample was then inoculated with 1 ml of a 24 h old Nutrient Broth culture of *C. joostei*, *C. gleum* and *C. indologenes* respectively and incubated at 25°C for 72 h. Samples were prepared in duplicate. Analytical procedures were as follows: The volatile compounds were determined just as to human. Each of the inoculated milk samples and control samples were measured into a 10 ml headspace clear glass-vial and closed with crimpable headspace caps with silicone-PTFE seals.

After incubation for 30 minutes at 70°C, a 1 ml headspace sample was withdrawn using a gas-tight syringe and injected into a Varian Chrompack CP- 3800 gas chromatograph equipped with a flame ionisation detector. A 30 m Zebron ZB-FFAP column, with a 25 µm film thickness and containing a nitroterephthalic acid modified polyethylene glycol liquid phase, with 40 to 250/260°C temperature limits was used. The temperature profile used was: 80°C for 2 min, increased by 5°C/ min to 250°C and held for 10 minutes. The total run time was 27 minutes. The carrier gas was hydrogen UHP. The column flow was 2.0 ml/min, the linear velocity 50.9 cm/ sec and the total flow 27.8 ml/min. Identification of the unknown compounds was achieved by comparing their retention times to those of analytical grade standard compounds. An ethylacetate stock solution was prepared as standard and contained 200 mg.l-1 of each of propionic acid, valeric acid, 1-Heptanol, 1-Octanol, 1-Nonanol, 1-Decanol, 1-Pentanol, 1-Hexanol, 2-Octanone, 2-Butanone, 2-Heptanone, acetone, 4-Methyl-2-pentanol, 2-Methyl-1-butanol, ethyl acetate, methyl acetate, propyl acetate and n-buthanol.

Sensory Analysis of Milk

Ten candidates for the semi-trained sensory odour descriptive panel were screened on the basis of their ability to differentiate among odours and are all scientists employed by, or postgraduate students in the Food Science Division. The panel consisted of seven female and three male persons, ranging from 23 to 60 years of age. As part of the training process, the panellists performed a paired comparison test and were asked to identify the spoiled sample by doing a sniffing test. The spoiled sample consisted of 10 ml fat free milk which was inoculated with *Chryseobacterium joostei* and incubated at 25°C for 72 h. The control sample was uninoculated fat free milk. For the sniffing test, 10 ml samples of Clover, UHT Process, Fat Free and Full Cream milk in McCartney bottles were each inoculated with 0.1 ml of a 24 h old Nutrient Broth culture of C. joostei, *C. gleum* and *C. indologenes* respectively and incubated at 25°C for 72 h.

An uninoculated milk sample of each type of milk was included to serve as controls. Thirty minutes before sensory evaluation, the milk samples were incubated at 50°C to allow better assessment of the volatile fraction. The milk samples were presented in the same McCartney bottles, in which the inoculations were done. These bottles had tight fitting caps to prevent any unnecessary odour release, which could influence results. Samples were coded with three digit numbers and served randomly on white trays. The testing was done under red lights to mask any colour differences which occurred due to the characteristic yellow to orange pigmentation produced by Chryseobacterium species. The inoculated samples, along with two control samples, one each of the fatfree and full cream milk, were presented to the panel in individual booths. Samples were uncapped, sniffed and removed immediately so, as not to cause exhaustion.

When exhaustion did occur, panellists were instructed to smell their own skin or clothing. As part of the training, panellists were not allowed to wear any perfumed personal care products or smoke for 30 minutes prior to sniffing, as this could influence the results. Three spoilage levels were chosen to define the intensity of the spoilage: level 1, no spoilage noted; level 2, weak spoilage noted; level 3, strong spoilage noted. Panellists also described the type of odour, if any, produced by each of the three cultures. The samples were regarded as weakly spoiled and strongly spoiled when at least 50 per cent of the panellists rated it at level 2 and level 3, respectively. The sensory testing was done in individual booths at the sensory facility of the Food Science Division, Department of Microbial, Biochemical and Food Biotechnology, University of the Free State.

FOOD BORNE BACTERIA

Harmful bacteria are the most common causes of food borne illnesses. Most of the bacteria that cause these illnesses go undetected because they

Introduction

have no odour and cause no change in the colour or the texture of the food that they are growing on. Therefore it would be fairly easy to cause an epidemic in a fairly large population using a bacteria to cause infection. Bioterrorism is a constant threat in the U.S. today and must be monitored in every way possible. Food is another way that bioterrorists could effect America. By infecting major foods consumed in the U.S. everyday an outbreak of illness could be implicated without much trouble. There are many bacteria that could be used to cause these illnesses, and some of the more effective ones are listed below telling what they are and the symptoms that their diseases entail. Food should always be handled very carefully even when not considering it from a bioterrorist point of view. Bacteria from food can come from many different places including: raw or undercooked meat, unwashed or contaminated produce, seafood, food handling, and cross contamination from other areas. Cross contamination can happen wherever the food was processed or packaged or in your own home when countertops and other areas where food was handled is not properly cleaned.

There are many preventative measures that can be taken to ensure not to ingest pathogenic bacteria and some of these are listed below:

- Make sure that food is promptly refrigerated. If prepared food is left in room temperature for more than two hours it may not be safe to eat. Make sure that your refrigerator is set at 40°F or lower and your freezer at 0°F or lower.
- Make sure that food is cooked at the appropriate temperature and for long enough or all of the bacteria will not be eliminated. (145°F for red meat, 160°F for pork and ground beef, 165°F for ground poultry and 180°F for whole poultry)
- Prevent cross contamination. Keep raw meat, poultry, seafood, and their juices away from other foods that are ready to eat.
- Ensure that food in handled properly by washing hands before and after preparation. Also make sure that surfaces where food in handled is always clean.
- Never defrost food on the kitchen counter.
- Do not pack the refrigerator. Cool air must circulate to keep food safe.

Everyone is always at risk when food borne bacteria are concerned so food must always be handled carefully but there are some that are at higher risk than others. Young children who have not built up a strong immune system yet are at higher risk along with pregnant women and their fetuses. Elderly people are also at a higher risk because of their lowered immunity. Another group are those that are immunodeficient due to cancer treatment, HIV, or the aids virus. These people should be especially careful and never eat any kind of raw food but all people should take the proper precautions.

GRAM POSITIVE RODS AND COCCI

Bacillus Cereus

Bacillus cereus is a gram-positive, aerobic spore former that causes an intoxication. Two types of toxins can be produced -- one results in diarrheal syndrome and the other in the emetic syndrome. Onset for the diarrheal syndrome is 6-15 hours after ingestion, with a duration of 24 hours. The primary symptom is diarrhea; vomiting is rare. Onset for the emetic syndrome is earlier -- 30 minutes to 6 hours after eating. As with the diarrheal syndrome, the duration is 24 hours. Food counts of B. cereus greater than 106/gram (1,000,000) indicate active growth and a potential hazard to health.

B. cereus is widely distributed throughout the environment. It has been isolated from a variety of foods, including meats, dairy products, vegetables, fish, and rice. The bacteria can also be found in starchy foods such as potato, pasta, and cheese products, and in food mixtures, such as sauces, puddings, soups, casseroles, pastries, and salads.

Fried rice is a leading cause of B. cereus emetic-type food poisoning in the U.S. The organism is frequently present in uncooked rice, and its heat-resistant spores survive cooking. If the rice is then held at room temperature, the spores might germinate and multiply. The toxin produced can survive heating (for instance, stir frying), and many people are unaware that cooked rice is a potentially hazardous food.

This organism will grow at temperatures as low as 39°F, at a pH as low as 4.3, and at salt concentrations as high as 18 per cent. Unlike other pathogens, it is an aerobe, and will grow only in the presence of oxygen. Both the spores and the emetic toxin are heat-resistant.

Listeria Monocytogenes

Unlike B. cereus, Listeria monocytogenes is not easily controlled by refrigeration. Listeriosis, the disease caused by this organism, can produce mild flu-like symptoms in healthy individuals. In susceptible individuals,

Introduction

including pregnant women, newborns, and the immunocompromised, the organism might enter the blood stream, resulting in septicemia. Ultimately, listeriosis can result in meningitis, encephalitis, spontaneous abortion, and stillbirth. The onset of disease might range from a few days to three weeks. The infectious dose is unknown.

L. monocytogenes can be isolated from soil, silage, and other environmental sources. It can also be found in man-made environments, such as food processing establishments. Generally, the drier the environment, the less likely the environmental will harbour this organism.

L. monocytogenes has been associated with raw or inadequately pasteurized milk, cheeses (especially soft-ripened types), ice cream, raw vegetables, fermented sausages, raw and cooked poultry, raw meats, and raw and smoked fish. In 1985, Mexican-style cheese led to at least 46 stillbirths in California. The consumption of large quantities of smoked mussels in New Zealand is reported to have caused two women to experience spontaneous abortions. The CDC has linked listeriosis with eating raw hot dogs or undercooked chicken.

L. monocytogenes is a psychotropic facultative anaerobe. It can survive some degree of thermal processing, but can be destroyed by cooking to an internal temperature of 158°F for two minutes. It can also grow at refrigerated temperatures below 31°F. Reportedly, it has a doubling time of 1.5 days at 40°F.

There is nothing unusual about Listeria's pH and water activity range for growth.

L. monocytogenes is salt-tolerant: it can grow in up to 10 per cent salt and has been known to survive in 30 per cent salt. It is also nitrite-tolerant.

Prevention of recontamination after cooking is a necessary control; even if the product has received thermal processing adequate to inactivate L. monocytogenes, the widespread nature of the organism provides the opportunity for recontamination. Furthermore, if the heat treatment has destroyed the competing microflora, L. monocytogenes might find itself in a suitable environment without competition.

Clostridium Perfringens

Clostridium perfringens is an anaerobic spore former and is a common cause of foodborne gastroenteritis. Perfringens poisoning is characterized by intense abdominal cramps and diarrhea, which begin 8-22 hours after eating contaminated food. The food must contain large numbers (100,000,000 or more) of the bacteria in order to produce toxin in the intestine.

The illness is usually over within 24 hours, but less severe symptoms might persist in some individuals for 1-2 weeks. A few deaths have been reported as a result of dehydration and other complications.

CDC estimates that there are 10,000 cases per year. Of these, approximately 1,200 are reported. The large number of cases and the small number of outbreaks are jointly attributable to institutional feeding, such as school cafeterias and nursing homes. Perfringens poisoning most frequently occurs in the young and older adults.

C. perfringens is widely distributed in the environment and is frequently found in the intestines of humans and many domestic and feral animals.

Spores of the organism persist in soil and sediments. C. perfringens has been found in beef, pork, lamb, chicken, turkey, stews, casseroles and gravy. In one 1984 outbreak involving 77 prison inmates, the implicated food was roast beef. Soon afterwards, there was a second outbreak which involved many of the same people, and on that occasion, the food implicated was ham. The cause in these instances was determined to have been inadequate refrigeration and insufficient reheating of the implicated foods.

In 1985, a large outbreak of C. perfringens gastroenteritis occurred among factory workers in Connecticut, and some 600 employees were affected. In that case, gravy that was prepared 12-24 hours before serving and inadequately cooled was implicated.

Clostridium perfringens is a mesophilic organism. Because it is also a spore-former, it is quite resistant to heat, and temperatures for growth range from 50°F to 125°F. The pH, water activity and salt ranges for growth are fairly typical.

Cooking the spores does not kill them. Cooking encourages them to germinate when the food reaches suitable temperature. Rapid, uniform cooling after cooking is critical. In virtually all outbreaks, the principal cause of perfringens poisoning is failure to properly refrigerate previously cooked foods, especially when it is prepared in large portions. Proper hot-holding (above 135°F) and adequate reheating of cooked, chilled foods (to a minimum internal temperature of 75°F) are also necessary controls. Educating food handlers remains a critical aspect of control.

Introduction 19

C. BOTULINUM

Like perfringens, C. botulinum is an anaerobic spore-former. There are seven types of C. botulinum -- A, B, C, D, E, F, and G -- but the types that will discussed be are type A, which represents a group called proteolytic botulinum, and type E, which represents the non-proteolytic group. The reason for the distinction is the proteolytic organisms' ability to breakdown protein.

This organism is one of the most lethal foodborne pathogens. The infectious dose is exceedingly low; a few nanograms of toxin can cause illness, and everyone is susceptible. Typically, the onset might be from 18-36 hours after eating contaminated food, but this can vary from 4 hours to 8 days. Symptoms include weakness and vertigo, followed by double vision and progressive difficulty in speaking, breathing, and swallowing. There might also be abdominal distention and constipation. The toxin eventually causes paralysis, which progresses symmetrically downward, starting with the eyes and face, and proceeding to the throat, chest, and extremities. When the diaphragm and chest muscles become involved, respiration is inhibited, and death from asphyxia results. Treatment includes early administration of antitoxin and mechanical breathing assistance. Mortality is high; without antitoxin, death is almost certain.

There is a variation of botulism known as infant botulism. In this case, the toxin is formed in the intestinal tract rather than preformed in the food. Honey is the only food that has been definitely linked to this disease, and it has occurred only in infants. The symptoms begin with constipation, followed by loss of appetite, lethargy, general weakness, pooled oral secretions, altered cry, and loss of head control, which is striking. Given the potential danger, one should never feed an infant honey.

C. botulinum is widely distributed in nature and can be found in soils, sediments from streams, lakes and coastal waters, the intestinal tracts of fish and mammals, and the gills and viscera of crabs and other shellfish. Type E is most prevalent in fresh water and marine environments, while type A is generally found terrestrially.

C. botulinum has been a problem in a wide variety of foods -- canned foods, acidified foods, smoked and uneviscerated fish, stuffed eggplant, garlic-in-oil, baked potatoes, sauteed onions, black bean dip, meat products, and marscapone cheese.

Two outbreaks in the 1960's involved vacuum-packaged fish (smoked ciscos and smoked chubs). The causative agent in each case was C. botulinum type E. The food was packed without nitrites, with low levels of salt, and were temperature-abused during distribution, all of which contributed to the formation of the toxin. There were no obvious signs of spoilage because aerobic spoilage microorganisms were inhibited by the vacuum packaging, and because type E does not produce any offensive odors.

In 1987, there were eight cases in which kapchunka -- an uneviscerated, salted, air-dried whitefish -- was implicated. It was believed that the fish contained low levels of salt during air drying at room temperature, which allowed for the toxin formation. The outbreak resulted in one death. Three cases of botulism in New York were traced to chopped garlic bottled in oil, which had been held at room temperature for several months before it was opened. Presumably, the oil created an anaerobic environment.

Type A and type E vary in their growth requirements. Minimum growth temperature for type A is 50°F, while type E will tolerate conditions down to 38°F. Type A's minimum water activity is 0.94, and type E's is -.97 -- a small difference but important when controlling the organism. The acid-tolerance of type A is reached at a pH of 4.6, while type E can grow at a pH of 5. And type A is more salt-tolerant; it can handle up to 10 per cent, while 5 per cent is sufficient to stop the growth of type E.

Although the vegetative cells are susceptible to heat, the spores are heat-resistant and able to survive many adverse environmental conditions. Type A and type E differ in the heat-resistance of their spores. Compared to type E, type A's resistance is relatively high. By contrast, the neurotoxin produced by C. botulinum is not resistant to heat, and can be inactivated by heating for 10 minutes at 176°F.

There are two primary strategies to control C. botulinum. The first is destruction of the spores by heat (thermal processing). The second is to alter the food to inhibit toxin production -- something that can be achieved by acidification, controlling water activity, adding salt and/or preservatives, and refrigeration. Water activity, salt, and pH can each be individually considered a full barrier to growth, but very often these single barriers -- a pH of 4.6 or 10 per cent salt -- are not used because they result in a product which is unacceptable to consumers. For this reason, multiple barriers are used.

Introduction

One example of a product using multiple barriers is pasteurized crabmeat stored under refrigeration. Type E is destroyed by the pasteurization process, while type A is controlled by refrigerated storage. (NOTE: Type E is more sensitive to heat, while type A's minimum growth temperature is 50°F.)

Another example of multiple barriers is hot-smoked, vacuum-packed fish. Vacuum-packaging provides the anaerobic environment necessary for the growth of C. botulinum, even as it inhibits the normal aerobic spoilage flora that would otherwise offer competition and exhibit signs of spoilage. So heat is used to weaken the spores of type E, which are then further controlled by the use of salt, sometimes in combination with nitrites. Finally, the spores of type A are controlled by refrigeration.

Vacuum-packaging of foods that are minimally processed, like sous vide foods, allows the survival of C. botulinum spores while wiping out competing microflora. If no control barriers are present, the C. botulinum might grow and produce toxin, particularly if there is temperature abuse. Given the frequency of temperature abuse documented at the retail and consumer levels, this process is safe only if temperatures are carefully controlled to below 38oF throughout distribution.

Vacuum-packing is also used to extend the shelf-life of the product. Because this provides additional time for toxin development, such food must be considered a high risk. Controls can be used to prevent the recurrence of such incidents as the 1987 outbreak caused by uneviserated fish. Any seafood product which will be preserved using salt, drying, pickling, or fermentation must be eviscerated prior to processing; the only exception is small fish (less than five inches in length), which will instead be processed to inhibit the formation of C. botulinum toxin -- something that can be done by maintaining a water phase salt of 10 per cent, a water activity of below 0.85, or a pH of 4.6 or less.

Staphylococcus. Aureus

Staphylococcus. aureus is a gram-positive cocci that grows in irregular clusters and produces a highly heat-stable toxin. Staphylococcal food poisoning is one of the most economically important foodborne diseases in the U.S., costing approximately $1.5 billion each year in medical expenses and loss of productivity. Onset is rapid, usually within four hours of ingestion, and the most common symptoms are nausea, vomiting, abdominal cramps, diarrhea, and prostration. Recovery usually takes two days, but can take longer in severe cases. S. aureus food poisoning is usually considered a mild, self-limiting illness with a low mortality rate; however, death has been known to occur among infants, older adults, and severely debilitated individuals. The infective dose is less than 1.0 microgram of toxin, and this toxin level is reached when the S. aureus population reaches 100,000 cells per gram in the food.

S. aureus can be found in air, dust, sewage, and water, although humans and animals are the primary reservoirs. S. aureus is present in and on the nasal passages, throats, hair and skin of at least one out of two healthy individuals. Food handlers are the main source of contamination, but food equipment and the environment itself can also be sources of the organism.

Foods associated with S. aureus include poultry, meat, salads, bakery products, sandwiches, and dairy products. Due to poor hygiene and temperature abuse, a number of outbreaks have been associated with cream-filled pastries and salads such as egg, chicken, tuna, potato, and macaroni.

S. aureus grows and produces toxin at the lowest water activity (0.85) of any food pathogen. Like type A botulinum and Listeria, S. aureus is salt-tolerant and will produce toxin at 10 per cent.

Foods that require considerable handling during preparation and that are kept at slightly elevated temperatures after preparation are frequently involved in staphylococcal food poisoning. And, while S. aureus does not compete well with the bacteria normally found in raw foods, it will grow both in cooked food and in salted food where the salt inhibits spoilage bacteria. Because S. aureus is a facultative anaerobe, reduced oxygen packaging can also give it a competitive advantage.

The best way to control S. aureus is to ensure proper employee hygiene and to minimize exposure to uncontrolled temperatures. While the organism can be killed by heat, the toxin cannot be destroyed even by thermal processing.

2

Foods from Microorganisms

Microorganisms are widely used in the food industry to produce various types of foods that are both nutritious and preserved from spoilage because of their acid content.

DAIRY FOODS

In the dairy industry, many products result from fermentation by microorganisms in milk and the products of milk. For example, buttermilk results from the souring of low-fat milk by lactic acid. The flavour is due to substances such as diacetyl and acetaldehyde, which are produced by species of Streptococcus, Leuconostoc, and Lactobacillus as they grow.

A fermented milk product with a puddinglike consistency is yogurt. Two bacteria, Streptococcus thermophilus and Lactobacillus bulgaricus, are essential to its production. After the milk has been heated to achieve evaporation, the bacteria are added, and the condensed milk is set aside at a warm temperature to produce the yogurt.

Sour cream is produced in a similar way, using cream as a starter material. The protein portion of the milk, the casein, is used to produce cheese and cheese products. Precipitated from the milk, the protein curd is an unripened cheese such as cottage cheese. The leftover liquid, the whey, can be used to make cheese foods. When the cheese is allowed to ripen through the activity of various microorganisms, various cheeses are produced.

Soft cheeses, such as Camembert, do not spoil rapidly. Camembert cheese is a product of the growth of the fungus Penicillium camemberti. Hard cheeses have less water and are ripened with bacteria or fungi. Swiss cheese is ripened by various bacteria, including species of Propionibacterium, which produces gas holes in the cheese. Bleu cheese is produced by Penicillium roqueforti, which produces veins within the cheese as it grows.

OTHER FERMENTED FOODS

Other fermented foods are also the product of microbial action. Sauerkraut, for example is produced by *Leuconostoc* and *Lactobacillus* species growing within shredded cabbage. Cucumbers are fermented by these same microorganisms to produce pickles.

Bread

Bread is still another product of microbial action. Flour, water, salt, and yeast are used to make the dough. The yeast most often used is *Saccharomyces cerevisiae*. This organism ferments the carbohydrates in the dough and produces carbon dioxide, which causes the dough to rise and creates the soft texture of bread. Unleavened bread is bread that contains no yeast. Sourdough bread can be made by using lactic acid bacteria to contribute a sour flavour to the dough.

WHY DO WE NEED TO PRESERVE FOODS

Now that you have understood the definition of food preservation, did you think about why we should preserve foods? Can you think of a few reasons for preserving foods? Let us find an answer to this question by taking an example of one food item. Let us take the example of a fruit, say mango. There are many ways by which mangoes can be preserved. These are: juice, murraba, squash, aam papad, pulp, chutney, pickle, raw mango powder.

Mango is a summer fruit and grows in large quantities during the months of April to August. Different varieties of mango are grown in different parts of our country. Usually all the quantity grown in a region cannot be consumed by the people staying there as there is always an excess. What does the farmer do with this excess quantity? He makes arrangements to transport the excess quantity to regions where either mango is not grown or where that particular variety of mango is not available. If he does not do this, the excess produce will rot and go waste. The farmer will then lose money. There is still some quantity which is left after the fresh fruit is consumed by the people. It is this quantity

Foods from Microorganisms

which has to be preserved for consumption during the months when mango is not available.

Preservation of foods is done during the months when food is available in large quantity and therefore at low cost:

- One of the important reasons for preserving foods is to take care of the excess produce. There are many other reasons for preserving foods. Let us learn about these.
- The second reason for preserving foods is that they add variety to our meals. Have you ever got tired of eating the same vegetables which are in season? Is it not nice to eat peas when they are either very expensive in the market or are not available ? Eating cauliflower in pulav or cauliflower vegetable during the summer months adds to the interest in meals. In the same way, eating some chatni, papad or pickle along with the meals adds to the variety. Preserving foods when they are in season makes this possible.
- Reaches areas where the food item is not grown. In some areas of Rajasthan which are desert areas and in Himalayan regions that are covered with snow most of the time, very few foods can be grown. Availability of some preserved foods can add to the variety and nutritive value of meals. For example inclusion of dehydrated peas, green leafy vegetables, canned fruits etc., in the meals is a good idea in such areas.
- Makes transportation and storage of foods easier. Preservation of foods usually reduces bulk. This makes their transportation and storage easier since it requires less space. For example, if you dry green leafy vegetables such as mint, methi, corriander, etc., their weight and volume reduces, thus making their storage easy.

Why Does Food Get Spoilt

The definition of food preservation states that preservation is keeping food in such a state that they do not get spoilt for a long period. Before we look at the reasons of food spoilage, let us understand, when is a food spoilt. When you keep bread outside the refrigerator for few days, a spongy growth is seen on it, which may be white, green on black in colour.

The bread thus gets spoilt due to growth of mould and becomes unfit for consumption. Likewise, if cooked dal or vegetables is left outside for sometime, it develops a bad smell and bubbles due to fermentation.

The dal and vegetables are thus spoilt and cannot be eaten. Can you now say when is a food spoilt? Food is said to be spoilt if there is rotting, *i.e.*, bad smell, fermentation, *i.e.*, bubbles/gas in the food or mould *i.e.*, spongy growth on the foodstuff. Formation of soft spots or soft brown spots on fruits and vegetables is also food spoilage. Why do foods get spoilt? If you know the reasons of food spoilage, you can remove these conditions while preserving food items. Foods get spoilt mainly due to the presence of micro-organisms, enzymes (present in foods), insects, worms, and rats.

Presence of Micro-organisms

Micro means small. Micro-organisms are very small organisms which cannot be easily seen. Micro-organisms spoil food items when the condition for their growth are appropriate. What are these appropriate conditions? Like all living beings micro-organisms require air, moisture, right temperature and food to grow and multiply.

The situations which provide appropriate conditions for growth of micro-organisms, can be listed as:

- Food having high moisture content.
- Air around the food containing micro-organisms.
- Foods kept for a long time at room temperature.
- Skin of fruits and vegetables getting damaged, thus exposing the food to micro-organisms.
- Foods with low salt, sugar or acid content.

Presence of Enzymes

Enzymes are chemical susbtances found in all plants and animals. Are enzymes harmful to foods? No, enzymes help in ripening of fruits and vegetables. A raw green mango after a few days becomes sweet in taste and yellow in colour due to the enzymes action. What will happen if you keep this yellow, ripe mango for a few more days?

It will become soft, develop black spots and will start smelling bad. This is due to continued action of enzymes. No one likes to eat such as over ripe, spoilt mango. You know that even when the skin of fruits is not cut or damaged, it gets spoilt. This is due to enzyme action.

Principles of Food Preservation

After learning about the causes of food spoilage, it should not be very difficult to list the principles of food preservation. Remember, a good method of food preservation is one that slows down or prevents altogether the action of the agents of spoilage. Also, during the process of food preservation, the food should not be damaged. In order to achieve this, certain basic methods were applied using the knowledge gained form observation of the effects of natural conditions on different types of foods. For example in earlier days, in very cold weather condition, ice was used to preserve foods. Thus, very low temperature became an efficient method for preventing food spoilage.

Let us now list the principles of food preservation:

- *Removal of Micro-organisms or Inactivating Them*: This is done by removing air, water (moisture), lowering or increasing temperature, increasing the concentration of salt or sugar or acid in foods. If you want to preserve green leafy vegetables, you have to remove the water from the leave so that micro-organisms cannot survive. You do this by drying the green leaves till all the moisture evapourates.
- *Inactivating Enzymes*: Enzymes found in foods can be inactivated by changing their conditions such as temperature and moisture, when you preserve peas, one of the methods of preservations is to put them for a few minutes in boiling water. This method inactivates enzymes and thus, in preserving the food.
- *Removal of Insects, Worms and Rats*: By storing foods in dry, air tight containers the insects, worms or rats are prevented from destroying it.

FOOD PRESERVATION WITH HOME CANNING HALTS SPOILAGE

Food preservation in canning jars is accomplished by killing spoilage causing agents with heat, removing air from the food products, and sealing the jars so that air and yeasts, moulds, and bacteria cannot be reintroduced to the food.

Four Causes of Food Spoilage

- Enzymes – Destroyed at 140°F.
- Moulds – Destroyed at 140°F to 190°F.
- Yeasts – Destroyed at 140°F to 190°F.
- Bacteria – Many types of bacteria exist. The toxins produced by some bacteria are also a hazard. Bacteria and associated toxins are destroyed in heat ranges from 190°F to 240°F.

Some bacteria are very tough and resist death even at high temperatures. The toughest bacterium is *Clostridium botulinum,* whose spores cause the deadly botulism. This bacterium is killed at 190°F and its toxic spores are destroyed at 240°F. This bacterium thrives in low-acid or non-acid foods in the absence of air. Foods in this category include corn, beans, peppers, poultry, fish, and meat. This is the main reason that these types of foods require the higher temperatures achieved during pressure canning. Boiling canned low-acid or non-acid foods for 10 to 20 minutes before eating them will destroy potential lingering toxins. This added step will give you one more reason to feel safe about eating home canned low-acid or non-acid foods.

Water Bath Canner

- *Equipment*: A large metal enamel kettle with lid and jar rack with the capacity to hold up to 7, quart-sized Mason jars. Widely available at discount stores and hardware stores. Cost: $20 to $25.
- *Uses*: The water bath canner is used to process Mason jars for home preservation of jams, fruits, fruit juices, and pickled vegetables.
- *Safety Considerations*: It is a large kettle of boiling water so be careful to avoid steam burns and splashing hot water.

Tested recipes are widely available from the USDA, numerous University agriculture departments, Mason jar manufacturers, fruit pectin manufacturers, and from friends and family who have recipes that are known to be safe. As you familiarize yourself with the principles of safe home canning, you will be able to judge all recipes for safety and even develop your own.

How to Use Water Bath Can

Please always review the manufacturer's directions for the Mason jars that you are using and pay close attention to the head space and processing times specified by recipes.

- Pick a recipe for the produce that you wish to can. Select only quality produce at or near the peak of ripeness. Prepare the recipe.
- While preparing the food, you will also need to get the Mason jars ready. Wash the jars, bands, and lids in hot soapy water. Only use new lids. Never re-use lids. Jars and bands can be re-used.
- Fill the water bath canner and start to heat it in order to sterilize your jars and lids. Use a large thermometer (a candy thermometer works well) to monitor the temperature. When the water is at least 180°F but less than boiling (212°F) add the jars and lids to the water.

Foods from Microorganisms

Sterilizing bands in unnecessary. Kerr and Ball lids specify that they should not be boiled, so make sure that the water stays just below the boiling point.
- Remove the jars and lids from hot water when the food is ready to be packed in jars.
- Add prepared food to jars. Use the handle of a wooden spoon along the insides of the jars to work out any air bubbles. Leave the specified head space, usually ¼ inch or ½ inch.
- Wipe the edges of jar mouths very carefully. They need to be completely clean. Any food particles or other debris on the mouth edge will interfere with the sealing of the lids.
- Place lids on jars and screw on the bands. Only screw them on hand tight. You don't need to twist hard.
- With a jar lifter, place the Mason jars into the hot water. Make sure that jars are not touching each other or the side of the kettle.
- Bring the water to a boil. Once the water is boiling, you can begin the timer for the processing time specified by the recipe. Adjust the heat source as necessary to keep the water boiling but to prevent it from boiling over.
- Once processing time is complete, shut off heat source, and use a jar lifter to remove jars from the canner. Set the jars on a cloth in a location free of drafts. Do not disturb the jars for 12 to 24 hours. You will likely hear the lids "pop" or "snap" down not long after removal, but the jars need to cool completely to make sure the seal is complete. (Processing times change with your land elevation. Consult charts that come with recipes and/or equipment.)

Pressure Canner for Home Canning

- *Equipment*: A pressure canner that has a sealed lid and a gauge that measures the pressure created by boiling water and steam. Under pressure, the steam will achieve temperatures of 240 to 250°F. Such high temperatures are necessary to destroy the bacterium *Clostridium botulinum*. Canners for home use typically cost between $80 and $130. Major brand is Presto.
- *Uses*: To process canning jars of fruit, vegetables, meat, fish, and poultry for purposes of preservation. Pressure canning is the only safe method for preserving low-acid vegetables and meats. Note, the

pressure canner can also be used like a pressure cooker to prepare meals.
- *Safety Considerations*: Proper use of a pressure canner requires diligent monitoring of the pressure gauge during operation and maintenance of the plugs, gaskets, and metal parts. An actual explosion of the equipment would only occur if the heat was left on and the pressure climbed into the danger zone.

Even so, the plug in the safety valve should blow out before a catastrophic failure of the pressure canner happened. Note that the danger zone for the equipment is several pounds of pressure beyond the pressures necessary for processing foods. To avoid problems, keep the pressure within the safe zone by watching the gauge and adjusting the heat source. In case of emergency, rapid cooling of the canner can be initiated by running it under cold water.

How to Use Pressure Can

Please follow the manufacturer's directions that accompany your specific pressure canner model. In general:
- The pressure canner will be filled with approximately 3 quarts of water (this amount would vary depending on size of canner).
- The water is brought to a boil and the prepared Mason jars are placed in the canner.
- The lid is placed on the canner, sealed, and locked, but the pressure regulator is *not* put in place yet.
- Once a free flow of steam is initiated through the vent, it will be allowed to vent for 10 minutes. (Time may vary depending on size of canner.) Venting steam exhausts air from the pressure chamber.
- After 10 minutes, the pressure regulator is put over the vent and pressure begins to build inside. When the interior becomes pressurized, the air vent/cover lock will rise and completely seal the chamber. Then pounds of pressure will start to accumulate and register on the gauge.
- Bring the canner to the pressure specified by the canning recipe and then maintain that pressure by adjusting heat source as necessary. You will find that once pressure has been achieved, the stove burner no longer needs to be on a high setting. Monitor the pressure closely to make sure it does not rise too far beyond desired pressure. Also, do not let the pressure fall below the pressure

Foods from Microorganisms

required by the recipe. It is important for the food within to be kept at the necessary pressure/temperature for the required amount of time.
- Once the food has processed at the required amount of time at the necessary pressure, turn off the heat source. Do not rapid cool the canner because this would cause jars within to break. (Processing times change with your land elevation. Consult charts that come with recipes and/or equipment.)
- Allow canner to cool until air vent/cover lock drops on its own. Then remove pressure regulator and allow canner to set for 10 more minutes.
- At this point you may unlock the lid and open it. Be careful of the steam that comes out because it will be scalding hot.
- With a jar lifter, remove Mason jars and set them on a cloth in a location free of drafts.
- Do not disturb jars for 12 to 24 hours after removal. They will be exceedingly hot. The contents will continue to boil after removal from the canner. As the jars cool, the lids will seal.

FOOD SAFETY FROM MICROORGANISMS

Microorganisms are small, living organisms including yeasts, molds, bacteria, and parasites. Some are beneficial and allow for the production of bread, cheese, wine and antibiotics. Others cause foods to spoil (e.g., mould) and yet others, called pathogens, make people sick by producing toxins or poisons. Microbes are found everywhere and can be moved to food by insects, utensils, equipment, hands, air, dust, and water.

The requirements for microbial life and growth are similar to that of humans-food, water, warmth, and in some cases, oxygen. They like the food we do and by growing in the food, may cause it to spoil or may cause human illness. In food processing, the goal is to take steps to: reduce microbial levels, remove microorganisms, and/or control their growth.

Bacteria are single-cell organisms that "grow" by dividing in two. Given the optimum conditions of nutrients, oxygen and temperature, a bacteria may divide every 7 to 20 minutes, thereby doubling the bacterial population. This means that if there is one cell present at first, there may be 512 cells present after 1 hour, 262,000 in two hours, and 4 million cells in 22 hours. As few as 500,000 cells of certain pathogens can cause food poisoning; 10,000,000 can cause food to spoil. Food processors play a crucial role in minimizing contamination of the product. Providing satisfactory cleaning procedures is important in controlling health hazards.

Definition of Terms

The following definitions are taken from the definitions and understandings of terms in section 201 of the Federal Food, Drug, and Cosmetic Act.

They are applicable to such terms when used in this part. The following definitions shall also apply:

- Acid foods or acidified foods means foods that have an balance pH of 4.6 or below.
- Satisfactory means that which is needed to complete the purpose of keeping with good public health practice.
- Batter means a semi-fluid substance, usually composed of flour and other ingredients, into which main components of food are dipped or with which they are coated, or which may be used directly to form bakery foods.
- Blanching, except for tree nuts and peanuts, means a prepackaging heat treatment of foodstuffs for enough time and at a satisfactory temperature to completely inactivate the naturally occurring enzymes and to affect other physical or biochemical changes in the food.
- Critical control point means a point in a food process where there is a high probability that unsafe control may cause, allow, or contribute to a hazard in the final food.
- Food means food that includes raw materials and ingredients.
- Food-contact surfaces are those surfaces that contact human food and those surfaces from which drainage onto the food or onto surfaces that contact the food ordinarily occurs during the normal course of operations. "Food-contact surfaces" include utensils and food-contact surfaces of equipment.
- Lot means the food produced during a period of time suggested by a specific code.
- Microorganism means yeasts molds, bacteria, and viruses. The term "undesirable microorganisms" includes those microorganisms that are of public health significance, that subject food to decomposition, that indicate that food is contaminated with filth,

or that otherwise may cause food to be contaminated. Occasionally in these regulations, FDA used the adjective "microbial" instead of using an adjectival phrase containing the word microorganism.
- Pest refers to any objectionable animals or insects including birds, rodents, flies, and larvae.
- Plant means the building used for the manufacturing, packaging, labeling, or holding of human food.
- Quality control operation means a planned and systematic procedure for taking all actions needed to prevent food from being contaminated.
- Rework means clean, uncontaminated food that has been removed from processing for reasons other than unsanitary conditions or that has been successfully reconditioned by reprocessing and that is suitable for use as food.
- Safe-moisture levels is a level of moisture low enough to prevent the growth of undesirable micro-organisms in the finished product under the conditions of manufacturing, storage, and distribution. The maximum safe moisture level for a food is based on its water activity (a_w). An aw will be considered safe for a food if there is satisfactory data available that shows the food at or below the given aw will not support the growth of undesirable microorganisms.
- Sanitize means to satisfactorily treat food-contact surfaces by a process that is effective in destroying vegetative cells of microorganisms of public health significance, and in substantially reducing numbers of other undesirable microorganisms, but without affecting the product or its safety for the consumer.
- Water activity (a_w) is a measure of the free moisture in a food and is the quotient of the water vapour pressure of the substance divided by the vapour pressure of pure water at the same temperature.

Foreign Material in Food Processing

Foreign material is anything that is not there naturally. For example, in potato processing plants, the only thing that should be on the line is potatoes. Everything else on the line is Foreign Material. Examples of foreign material found on the lines include: rocks, bottles, glass, metal and parts of cans, corn cobs, tools, hats, knives, jewelry, plastic, hair nets, wood and gloves.

As the potatoes become more processed, removing foreign material from the product becomes more important. Of all the foreign material, the most hazardous is Glass. The reasons for this are that not only is glass hard enough to break or jam machinery, but glass is also a material that will not be picked up by metal detectors. In addition, people eating the product can break teeth or have other medical problems because of glass.

The goal of every food processor is to identify possible hazards associated with their product, processing and distribution, which would compromise the safety and quality of the food. Processors must control hazards that pose a health risk to the consumer. Hazards in food can be:

- Chemical.
- Physical.
- Microbiological.

They can enter the food at any stage, from growing and harvesting, through to the point of purchase or consumption. Some ways to control hazards include: selecting ingredients from fields where agricultural chemical use is controlled and from waters where industrial pollution and hazardous microorganisms are known to be below maximum allowable limits. [Safe use of chemical food additives, cleaning chemicals; labeling products containing nuts; and avoiding the use of moldy ingredients.] Safe ingredient selection and thorough sanitation and hygiene, in processing and packaging, can control microbial contamination. The nature of microbes, and the widespread effects of microbial contamination, makes controlling these hazards very challenging to food processors.

Workers should Wear Clean Cloths

People working in the fields may not have to wear clean clothing all the time. However, people working with food items must wear clean clothing each day. Clean clothing helps keep the product clean. Clothing may look clean, but if it is dirty or has not been washed for a long time, then it is probably full of bacteria or germs that could affect the product. Try to keep your clothes clean. If they get dirty, be sure to take then home to wash them. Remember that we're working around food. Since you are working with food items, you are required to wear clean clothing each day. You may have an apron to wear to protect you and your clothes.

Be sure to keep it clean. Do not wear loose clothing around machinery or in the belts. Do not wipe your Hands on your clothing to dry them

Foods from Microorganisms

because this is not a clean practice. The Food and Drug Administration states what must be done in a food processing facility in order to meet the requirements for sanitation. These are the conditions that you would be helping to fulfill. while working in a food processing plant.

Sanitary Operations

- *General Maintenance*: Buildings, fixtures, and other physical facilities of the plant shall be continued in a sanitary condition and shall be kept in repair satisfactory to prevent food from becoming contaminated. Cleaning of utensils and equipment shall be conducted in a way that protects against contamination of food, food-contact surfaces, or food-packaging materials.
- *Substances Used in Cleaning and Storage of Toxic Materials*:
  - Agents used in cleaning procedures shall be free from undesirable microorganisms and shall be safe and satisfactory under the conditions of use. Compliance with this requirement may be achieved by any effective means including purchase of these substances under a supplier's guarantee or certification, or examination of these substances for contamination. Only the following toxic materials may be used or stored in a plant where food is processed or exposed:
    (a) Those required to provide clean and sanitary conditions;
    (b) Those needed for use in labouratory testing procedures;
    (c) Those needed for plant and equipment maintenance and operation;
    (d) Those needed for use in the plant's operations.
  - Toxic cleaning compounds, sanitizing agents, and pesticide chemicals shall be identified, held, and stored in a way that protects against contamina-tion of food, food-contact surfaces, or food-packaging materials. All relevant regulations published by other Federal, State, and local government agencies for the application, use, or holding of these products should be followed.
- *Pest Control*: No pests shall be allowed in any area of a food plant. Guard or guide dogs may be allowed in some areas of a plant if the presence of the dogs is unlikely to result in contamination of food, food-contact surfaces, or food-packaging materials. Effective steps shall be taken to exclude pests from the processing areas and to protect against the contamination of food on the premises by pests. The use of insecticides or rodenticides is allowed only under

precautions and restrictions that will protect against the contamination of food-contact surfaces, and food-packaging materials.

- *Sanitation of Food-Contact Surfaces*: All food-contact surfaces, including utensils and food-contact surfaces of equipment, shall be cleaned as frequently as needed to protect against contamination of food.

Sanitary Facilities and Controls

Each plant shall be equipped with satisfactory sanitary facilities and accommodations.

1. *Water Supply:* The water supply shall be satisfactory for the operations and shall be derived from a satisfactory source. Any water that contacts food or food-contact surfaces shall be clean. Running water at a suitable temperature, and under pressure as needed, shall be provided in all areas where required for the processing of food, for the cleaning of equipment, utensils, and food-packaging materials or for employee sanitary facilities.

- *Plumbing*: Plumbing shall be of satisfactory size and design and satisfactory installation and continue to:
  – Carry satisfactory quantities of water to required locations throughout the plant.
  – Safely convey sewage and liquid disposable waste from the plant.
  – Avoid constituting a source of contamination to food, water supplies, equipment, or utensils or creating an unclean condition.
  – Provide satisfactory floor drainage in all areas where floors are subject to flooding-type cleaning or where normal operations release or discharge water or other liquid waste on the floor.
  – Provided that there is not back-flow from, or cross-connection between, piping systems that discharge waste water or sewage and piping systems that carry water for food or food manufacturing.

- Sewage disposal shall be accomplished by a satisfactory sewage system or disposed of through other satisfactory means.

- *Toilet Facilities:* Each plant shall provide its employees with satisfactory, easily accessed toilet facilities. This requirement may be accomplished by:

Foods from Microorganisms

- Providing the facilities in a sanitary condition.
- Keeping the facilities in good repair at all times.
- Providing self-closing doors.
- Providing doors that do not open into areas where food is exposed to airborne contamination, except where alternate means have been taken to protect against such contamination (such as double doors or positive airflow systems).

• *Hand-washing Facilities:* Hand-washing facilities shall be satisfactory and convenient and be furnished with running water at a suitable temperature. This requirement may be accomplished by providing:
  - Hand-washing and, where appropriate, hand-sanitizing facilities at each location in the plant where good sanitary practices require employees to wash and/or sanitize their hands.
  - Effective hand-cleaning preparations.
  - Sanitary towel service or suitable drying devices.
  - Devices or fixtures, such as water control valves, so made and constructed to protect against re-contamination of clean hands.
  - Readily understandable signs directing emplo-yees handling unprotected food, unprotected food-packaging materials, of food-contact surfaces to wash their hands before they start work, after each break from post of duty, and when their hands may have become contaminated. These signs may be posted in the processing room(s) and in all other areas where employees may handle such food, materials, or surfaces.
  - Garbage receptacles that are built and maintained in a way that protects against contamination of food.

• *Rubbish and Offal Disposal:* Rubbish and any offal shall be so conveyed, stored, and disposed of as to minimize the development of odour, minimize the potential for the waste becoming an attractant and harborage or breeding place for pests, and protect against contamination of food, food-contact surfaces, water supplies, and ground surfaces.

When working in a sanitation position, employees will:
• Handle and use chemicals in a safe and effective way.
• Clean areas and equipment, using cleaning utensils and chemicals.
• Operate equipment, and be able to follow lockout/tagout procedures.

- Wear personal protective equipment that may include a wet suit, goggles, gloves, boots, and other equipment. Sanitation is an important part of making sure that microorganisms are controlled and food is safely produced. Individuals in sanitation positions need to understand the purpose and operation of the sanitation function, and make decisions based on sound judgment when a situation dictates.

MICROBIAL DETERIORATION OF FOOD COMPONENTS

The type and extent of microbial colonization of a food only partly affects its ultimate deterioration, because the biochemical activities of the microbial community structure at the time of the onset of spoilage are also decisive. Organoleptic deterioration may, however, occur before any marked chemical changes take place in the food. This is because some odiferous metabolites can be detected organoleptically at very low levels.

Less than 1 ppm dimethyl sulphide or methyl mercaptan is sufficient to cause off-odours. Even at the maximum cell concentration usually achieved, metabolising at the optimum rate would only produce about 2 ml g-1 h-1 of carbon dioxide. At lower temperatures this rate would be much less. Conversely, high levels of microbes may be present in a food that shows no obvious organoleptic change. The growth of microbes in foods inevitably causes chemical changes. Bacteria, the predominating organisms in the microbial ecology of most foods, are extremely small: a rod of 2×0.8 µm has a volume of about 10-12 cm^3. Although they have a high metabolic potential per cell, large numbers of bacteria are required before they can cause measurable chemical changes.

3

Food Safety

INTRODUCTION

Microorganisms are tiny, mostly one-celled organisms capable of rapid reproduction under proper growth conditions. Those microorganisms important in the food industry include the *bacteria, viruses, yeasts, molds*, and *protozoans*. Many are helpful and serve useful functions such as causing breads to rise, fermenting sugars to alcohol, assisting in the production of cheese from milk, and decaying organic matter to replenish nutrients in the soil. Microorganisms can also cause foods to spoil and make them inedible. Spoilage organisms cost the food industry millions of dollars each year. Microorganisms can also be harmful.

These are called *pathogens* and cause between 24 to 81 million cases of foodborne illness in the U.S. each year. These forms of life, some so small that 25,000 of them placed end to end would not span one inch, were little known until the last century. Antony van Leeuwenhoek and others discovered "very little animalcules' in rain water viewed through crude microscopes. We now know that microorganisms occur everywhere on the skin, in the air, in the soil, and on nearly all objects. It was not until Pasteur proved that microorganisms could be eliminated from a system, such as a can of food, and sealed out (*hermetically sealed*), that man could exert control over the microbes in his environment.

Terminology

Bacteria are single-celled microorganisms found in nearly all natural environments. Outward appearances of the cell such as size, shape, and arrangement are referred to as *morphology*. Morphological types are grouped into the general categories of spherical (the cocci), cylindrical (the rods) and spiral. The cocci may be further grouped by their tendencies to cluster. Diplococci attach in pairs, streptococci in chains, staphylococci bunch like grapes, and sarcinae produce a cuboidal arrangement.

Bacterial cells have definite characteristic structures such as the *cell wall, cytoplasm*, and nuclear structures. Some also possess hairlike appendages for mobility called *flagella, fimbriae* which aid in attachment, plus *cytoplasmic* and *membranous inclusions* for regulating life processes. *Viruses* are extremely small parasites. They require living cells of plants, animals, or bacteria for growth. The virus is mainly a packet of genetic material which must be reproduced by the host. *Yeast and mold* are *fungi* which do not contain chlorophyls. They range in size from single-celled organisms to large mushrooms. Although some are multi celled, they are not differentiated into roots, stems and leaves. The true fungi produce masses of filamentous *hyphae* which form the *mycelium*. Depending on the organism, they may reproduce by fission, by budding as in the case of yeasts, or by means of *spores* borne on fruiting structures depending on the organism.

Protozoa are single-celled organisms such as the amoeba which can cause disease in humans and animals. They possess cell structure similar to higher, more complex organisms. Microorganisms are referred to by their scientific names which are often very descriptive. The first part of the name, the genus, is capitalized such as *Streptococcus*, spherical cells which occur in strips, *Lactobacillus* which are rod-shaped organisms commonly found in milk, or *Pediococcus* spherical cells which ferment pickles. The second part of the name is not capitalized and gives added information. Both parts of the name are underlined or italicized as in the case of *Saccharomyces cerevisiae,* a yeast which commonly ferments sausage.

The Cell

The cell is the basic unit of life. Our bodies are made up of millions of cells, but many microorganisms are single celled creatures. Cells are basically packages of living matter surrounded by membranes or walls. Within the cell are various organelles which control life processes for the cell such as intake of nutrients, production of energy, discharge of waste materials, and reproduction. Growth of the cell normally means reproduction. Bacteria and similar organisms reproduce by *binary fission*, a splitting of a single cell into two. The control center for the bacterial cell is the *nuclear structure*.

Food Safety

Within it is the genetic material which is duplicated and transferred to daughter cells during reproduction. These daughter cells can again divide to produce four cells from the original one. The time It takes for a new cell to produce a new generation of daughter cells is called *generation time*.

Under optimum growth conditions, certain organisms can have a generation time of 15 minutes. In four hours over 65,000 cells could be produced from a single microorganism!

Under adverse conditions, certain bacteria can protect the cell's genetic material by producing *spores*. These are extremely resistant capsules of genetic materials. Though there are no discernible life processes in the spore, under proper sporulation conditions, a viable, reproducing cell will germinate from it.

Factors Affecting Growth

Microorganisms, like other living organisms, are dependent on their environment to provide for their basic needs. Adverse conditions can alter their growth rate or kill them.

Growth of microorganisms can be manipulated by controlling:
- Nutrients available
- Oxygen
- Water
- Temperature
- Acidity and pH
- Light
- Chemicals

Nutrients

Nutrients such as carbohydrates, fats, proteins, vitamins, minerals and water, required by, man are also needed by microorganisms to grow. Microbes differ in their abilities to use *substrates* as nutrient sources. Their enzyme systems are made available according to their genetic code. They vary in ability to use nitrogen sources to produce amino acids and, therefore, proteins. Some require amino acids to be supplied by the substrate. When organisms need special materials provided by their environment, we refer to them as *fastidious*. Difference in the utilization of nutrients and the waste products they produce are important in differentiating between organisms.

Oxygen

Microbes also differ in their needs for free oxygen. *Aerobic* organisms must grow in the presence of free oxygen and *anaerobic* organisms must grow in the absence of free oxygen. *Facultative* organisms can grow with or without oxygen, while *microaerophilic* organisms grow in the presence of small quantities of oxygen.

Water

Water is necessary for microbes to grow, but microbes cannot grow in pure water. Some water is not available. A measurement of the availability of water is aw or *water activity*. The aw of pure water is 1.0 while that of a saturated salt solution is 0.75. Most spoilage bacteria require a minimum aw of 0.90. Some bacteria can tolerate an aw above 0.75 as can some yeasts and most molds. Most yeasts require 0.87 water activity. An aw of 0.85 or less suppresses the growth of organisms of public health significance.

Temperature

Microorganisms can grow in a wide range of temperatures. Since they depend on water as a solvent for nutrients, frozen water or boiling water inhibits their growth. General terms are applied to organisms based on their growth at different temperatures. Most organisms grow best at or near room and body temperature. These are *mesophiles*. Those growing above 400C (1050F) are called *thermophiles* while those growing below 250C(750F) are called psychrotrophs.

Acidity

The nature of a solution based on its acidity or alkalinity is described as pH. The pH scale ranges from 0, strongly *acidic*, to 14, strongly *basic*. *Neutral* solutions are pH 7, the pH of pure water. Most bacteria require near neutral conditions for optimal growth with minimums and maximums between 4 and 9. Many organisms change the pH of their substrate by producing by-products during growth. They can change conditions such that the environment can no longer support their growth. Yeasts and molds are more tolerant of lower pH than the bacteria and may outgrow them under those conditions.

Light & Chemicals

Ultraviolet light and the presence of chemical inhibitors may also affect the growth of organisms. Many treatments such as hydrogen peroxide and chlorine can kill or injure microbes. Under certain conditions those given a sublethal treatment are injured, but can recover.

Growth

Characteristic growth patterns can be illustrated on a graph. There is a selected portion of the normal growth curve which is referred to as the *logarithmic growth phase* or the *log phase*. When cells begin to grow, we usually observe a period of no apparent growth which we refer to as the *lag phase*. This occurs because cells are making necessary adjustments to adapt.

Next we experience the rapid growth or the *log phase* previously described. As cell mass becomes large, nutrients are exhausted and metabolic byproducts collect. Growth tapers off and the population remains constant for a time. This is referred to as the *stationary phase* of growth. With no intervention in the system the population will enter a *death phase* and total numbers of organisms will decline.

Enumeration of Cells

Numbers of microorganisms can be estimated based on cell counts, cell mass, or activity. A *direct count* of cells may be made by examination of a known volume of cell suspension under a microscope. This method is rapid and requires minimal equipment. This does not distinguish living cells from dead and may be tedious.

The application of certain stains make visible morphological characteristics of the organism which can aid in identification. Probably the most common method of enumerating cells is the *plate count*. A known volume of a diluted specimen is added to agar in a petri dish. Assuming dilute solution and that each organism will divide until it develops a visible mass or *colony*, the colonies can be counted and multiplied by the dilution factor to estimate the number of organisms in the original sample.

This method is based on the assumption that a colony forming unit (CFU) is a single organism. This will not hold exactly true in the case of strips or clumps of cells. Sublethally injured organisms may not grow. The culture conditions may not be conducive to the growth of certain types of *fastidious organisms* such as anaerobes. To measure the progress of a culture in a clear broth, changes in *turbidity* can be measured and related to numbers or organism. This method is easy and rapid. Many commercial establishments have found application for this method.

Identification

As many types of cells look similar in morphology and produce similar colonies, it becomes necessary to identify the organisms by their biochemical characteristics. Biochemical testing requires pure cultures isolated from a single colony from a plate count or streaked plate made for isolation purposes. *Isolates* are grown in an enriched broth to produce large cell numbers. Various media can then be inoculated with the culture and then growth can be observed by carefully formulating the various media, the biochemical and growth characteristics of the organism can be determined. Previously determined morphological characteristics can be combined with biochemical data to properly classify the organism. Newer methods more rapidly identify organisms of interest using other characterization such as monoclonal antibodies and DNA.

THERMAL PROCESSING OF FOODS

Low-acid canned foods are regulated by 21CFR113. These foods have a pH of greater than 4.6 and have a aw greater than 0.85. the regulations require that a *scheduled process* established by a processing authority be selected by the manufacturer which renders the product, under the specified conditions, *commercially sterile*. Commercial sterility is determined by processing food inoculated with known quantities of microbial spores. The test organisms used should simulate the resistance of *Clostridium botulinum* under those conditions. A process which eliminates these spores will destroy *Clostridium botulinum* spores. Acid and acidified foods will not allow the growth of *Clostridium botulinum*. However, a *pasteurizing* heat treatment is necessary to destroy other bacteria, viruses, yeasts, and mold. Temperatures and process times which destroy microorganisms without destruction of nutrients may fail to deactivate enzymes. If the process selected does not inactivate the enzymes, product changes may proceed during storage, at an accelerated rate, and cause a loss of product quality. The microbiological quality of raw product is the simple greatest determinant in the level of quality in the food. Thermal processing is no substitute for good raw product quality.

The Retail Food Safety Need

As more retail food operations across the U.S. and throughout the world compete to feed consumers, it becomes essential that uniform hazard analysis and control guidelines for producing, buying, and selling food products be developed. These guidelines must be based on science and validated in actual operation. At this time, consumers in the U.S. are doing less food preparation themselves and are dining out and/or are relying on retail food outlets for ready-prepared items. Food operations, as defined in this document, include: food markets where food is sold to be prepared in the home; food preparation and foodservice establishments that include restaurants, institutional foodservice units, street vending operations, hotel and lodging operations, military commissaries; and even the home, which is actually a miniature foodservice unit.

Food science and technology have improved the understanding of the potential microbiological, chemical, and physical hazards in foods. This knowledge can be used to determine the criteria necessary to assure that food products and commodities meet consumer safety expectations with an acceptable risk at the raw material level, the distributor level, and the consumer level. International trade and tourism will be enhanced throughout the world when there is a clearer understanding between the producer, retailer/supplier, and the buyer of food concerning the potential hazards in food and the level of risk associated with consuming a food.

Beginning with *Codex Alimentarius* (32) and the International Commission on Microbiological Specifications for Food (ICMSF), and continuing with the National Advisory Committee on Microbiological Criteria for Foods (135), there has been a movement for many years for more complete safety specifications for foods in local, national and international trade. The result is the current emphasis on Hazard Analysis and Critical Control Points (HACCP) in food production facilities and retail food operations. However, people have lost sight of the fact that HACCP is only a part of a company's food production quality management program. A company cannot accomplish process hazard control until it has process quality control. Hazards and critical control points can be easily identified. However, it is a separate issue to actually operate so that there is a very low chance of process deviation and low risk of a hazardous item being produced.

HAZARD CLASSIFICATION

The World Health Organization (WHO) Division of Food and Nutrition (200) identifies a hazard as a biological, chemical, or physical agent or condition in food with the potential to cause harm. In addition to these factors, nutrient levels inadequate to prevent deficiency disease must also be considered as a hazard. Note that one food, or even a single serving, while it could cause foodborne illness, no retail dish or retailer (except in institutions) is responsible for balancing the client's entire diet.

Actually, food is never risk free. The Environmental Protection Agency (EPA) has developed many risk criteria for water and chemicals. Unfortunately, the consumer's standard for purchased food is zero risk, which is unattainable. There will never be totally pathogen-free food animals or fruits and vegetables, nor will there ever be human populations who excrete totally pathogen-free fecal material. Therefore, hazards must be assessed, and risks must be reduced to a safe level.

Hazard analysis is the process of collecting and interpreting information to assess the risk and severity of potential hazards (ICMSF, 90). It is appropriate to classify hazards as critical concern, major concern, and minor concern.

The definitions for hazards in food are as follows:
- *Critical Concern*: Without control, there is life-threatening risk.
- *Major Concern*: A threat that must be controlled but is not life threatening and requires no government intervention.
- *Minor Concern*: No threat to the consumer (normally quality and cleanliness issues).

For example, spoilage bacteria, even at more than 50,000,000 per gram, are of no known safety concern. Coliform bacteria include both spoilage and pathogenic microorganisms. Therefore, a coliform count of 1,000 CFU per gram is of minor concern until specified levels of specific pathogens in the coliform group are established. Only pathogens and pathogenic substances ingested above threshold levels can cause illness, disease, and death. Properly controlled levels of salt, sugar, and MSG are of no concern. Floors, walls, ceilings, and many other items grouped under Good Manufacturing Practices (31) are really minor concern.

Of minor concern is also the presence of 1,000 *Staphylococcus aureus* cells, *Bacillus cereus* spores, or *Clostridium perfringens* spores per gram of food. These organisms are not hazardous until they reach 100,000 vegetative cells per gram. A hazard of critical concern, on the other hand, is a dose of 1 or more *Escherichia coli* O157:H7, 100 *Salmonella* spp., or 500 *Campylobacter jejuni* in a portion of hamburger or chicken. It is essential that food be prepared by a cook who is trained to reduce potential hazards to a safe level. Even healthy people can become ill if they consume food containing these pathogens at high enough levels.

Food Safety

Some people develop natural immunity. People who live in an environment with greater levels of pathogenic agents have a greater tolerance. The example is farmers who acquire a natural immunity and elevated resistance to some illnesses after being exposed to many of these pathogens while working on their farms. It is the job of the cook to make food safe by washing the food, such as raw fruits and vegetables and by pasteurizing raw food with the application of heat. During preparation, care must be taken to prevent cross-contamination of any ready-to-eat foods when raw foods are handled. (Raw foods include fruits and vegetables as well as meat, fish, and poultry products.)

The Process of Hazard Identification

In order to develop a process or operation capable of protecting public health, which, in turn, will minimize liability costs, it is essential that a logical process for hazard identification be followed.

The following criteria should be included in the hazard analysis:

1. Evidence that the microbial, chemical, or physical agent is a hazard to health based on epidemiological data or a laboratory hazard analysis.
2. The type and kind of the natural and commonly acquired microflora of the ingredient or food, and the ability of the food to support microbial growth.
3. The effect of processing on the microflora of the food.
4. The potential for microbial contamination (or recontamination) and/or growth during processing, handling, storage, and distribution.
5. The types of consumers at risk.
6. The state in which food is distributed (e.g., frozen, refrigerated, heat processed, etc.).
7. Potential for abuse at the consumer level.
8. The existence of Good Manufacturing Practices (GMPs).
9. The manner in which the food is prepared for ultimate consumption (i.e., heated or not).
10. Reliability of methods available to detect and/or quantify the microorganism(s) and toxin(s) of concern.
11. The costs/benefits associated with the application of items 1 through 10 (as listed above).

Coupling hazard analysis with correct hazard control and operating procedures enables the retail food operator to demonstrate a high degree of "due diligence" in the prevention of problems. Due diligence is essential if an operator is to avoid punitive legal damages resulting from a lawsuit because he/she did nothing to control the hazards in the food.

Food-related Illness and Death in the United States

While cooks can control most pathogenic agents in food most of the time, there are still an enormous number of illnesses and deaths that occur each year in the U. S because of foodborne agents. More than 200 known diseases are transmitted through food. The causes of foodborne illness and disease include viruses, bacteria, parasites, toxins, metals, and prions, and the symptoms of foodborne illness vary from mild gastroenteritis to life-threatening neurologic, hepatic, and renal disorders.

A 1999 Centers for Disease Control and Prevention report (203) estimated cases of illness and deaths for the general population in the U.S. due to known causes. It can be seen that not all of these pasthogens are transferred by food. They are sometimes spread by water, person-to-person contact, or other means. Also note that 67% of the estimated 13.8 million foodborne illnesses of known etiology are viral (Norwalk-like viruses).

The authors of the 1999 CDC report (203) applied another analysis, taking into account under-reporting factors, to arrive at an estimate that predicts a much larger annual illness incidence of 76 million (mostly diarrheal illnesses of 1-to-2-day duration), 325,000 hospitalizations, and 5,000 deaths. This estimate is based on the authors' further speculation that 80% of the illnesses that occur annually are due to "unidentified etiological agents". The CDC report (203) did not estimate causes of illness or injury due to chemicals, toxins, or hard foreign objects in food. Predicted annual occurrences cited in Table below are those estimated by Todd (186).

The pathogens responsible for foodborne illness and disease are ubiquitous. Because of environmental and animal contamination, food and food products will always be contaminated with low levels of pathogens, and occasionally at high levels (especially poultry). At low levels, pathogenic microorganisms in food cause no problems, and people even develop immunity to their presence in food (74, 85, 91, 137, 151, 181, 198). At illness thresholds, however, pathogens in food can make people ill and cause death. Pathogens in food can only be controlled when food producers, food retailers, and consumers know the potential hazards in food, and handle and prepare food by methods to reduce the hazard to a tolerable level.

How do pathogens in food get to high levels? It can happen anywhere in the food chain, from growing the animal, fish, vegetable or fruit to consumption. Everyone in the food chain must have knowledge of the causes of foodborne disease and illness, and must establish a program that assures safety before they produce and sell food. If not, incomplete hazard control processes are implemented.

Government Microbiological Standards for Raw and Pasteurized Food

The Code of Federal Regulations has been interpreted by the United States Department of Agriculture and the Food and Drug Administration to mean that if a sample of a processed food is found to be contaminated with *E. coli* O157:H7, *Salmonella* spp., or *Listeria monocytogenes*, the food is deemed unfit for human consumption. Sample size is variable, often negative in 25 grams, but the products' microbial standard is not enforced in retail operations, because retail inspectors seldom check. In retail food operations, the presence of *L. monocytogenes* is likely on fresh produce, meat, fish, and poultry, as well as on floors and in floor drains. It has been estimated by Farber that raw food (e.g., coleslaw) can contain 100 to 10,000 CFU/g of *L. monocytogenes*. Meat products will have low levels of *E. coli* O157:H7 and *Salmonella* spp., and poultry is often found to be contaminated with *Campylobacter jejuni* and *Salmonella* spp.

Need for International Safe and Hazardous Level Guidelines: At the present time, there are few worldwide food safety guidelines for upper or lower control limits of potential hazards in foods, or the process limits. This document proposes limits. When standards do exist, they may be inappropriate (e.g., specification of the numbers of coliforms in milk and in shellfish waters), or they may be unattainable (e.g., a zero level of both *Salmonella* spp. and *L. monocytogenes* in food). There is no zero level in food safety. There is a point at which measurements cannot be made with any degree of statistical reliability, and this point is frequently taken as "zero." For example, processed food is assumed to be safe from salmonellae contamination if there are no detectable salmonellae in a sample using the method of analysis described by the Bacteriological Analytical Manual (54). However, as laboratory methods improve, standards for safe levels of pathogenic material in food may be established. When safety standards or guidelines are developed for microbiological, chemical, and physical hazards in food, the standards or guidelines must be based on the risk of causing injury or illness to consumers, not what the processing industry is capable of achieving, or what scientific technology is capable of measuring.

CONTAMINATION LEVELS AND MICROBIOLOGICAL CONTROL

Pathogen Contamination from Human Sources

In addition to the contamination of raw food supplies that occurs during growing, shipping, and processing, there is the problem of food contamination caused by people who are carriers of pathogens. While most food codes require that when an employee is sick, he or she should stay home, people actually shed pathogenic bacteria a few hours to many days before they have major symptoms of illness.

Food workers can become permanent carriers of pathogens and yet exhibit no signs of illness (e.g., *Salmonella* carriers). Therefore, the only safe assumption is that all employees who work with food every day carry pathogens (on their skin and in their urine and feces) that must be kept out of the food. The following table is a list of some pathogens that commonly originate from human sources.

Foodborne Illness Hazards: Threshold and Quality Levels

The next step in the systematic approach to hazard control is to recognize that certain pathogens can be selected as the basis for process control criteria. Table shows the levels of microorganisms that caused illness, based on volunteer feeding tests of healthy people. Hazardous levels of chemicals and hard foreign objects in food are also listed.

Spores of the pathogenic bacteria *B. cereus, C. botulinum,* and *C. perfringens* must germinate after the food is cooked, and the vegetative cells must multiply to a level of at least 10^4 to 10^6 in order to make the food a hazard to consumers.

Low levels of these spore-forming organisms (10^2 to 10^3) in food are not a hazard, with the exception of *C. botulinum* found in honey. Honey must not be fed to infants at an age of less than one year (103). Infants have few competitive gut microflora, and the *C. botulinum* in the honey is able to germinate and then multiply in the infant's intestinal tract and produce toxin at a level sufficient to cause illness and even death. This hazard can be controlled by educating family members not to feed honey to babies.

Campylobacter jejuni is a principal cross-contamination problem. This pathogenic microorganism may be present in poultry at a "natural" level after slaughter that will cause illness without having to multiply. Raw chicken and other poultry can be contaminated at high levels (i.e., 10^6 to 10^7 organisms / chicken) (84). The threshold for illness in healthy people is approximately 500 organisms in 180 ml of milk, which means approximately 2 organisms per ml. Comparing this infective level with those for *Salmonella*, *E. coli*, *Vibrio*, etc., it is obvious that *C. jejuni*, in poultry that is grossly contaminated, will be a major cause of foodborne illness.

Other human fecal pathogens of concern include *Shigella* spp., pathogenic *E. coli*, *Salmonella* spp., hepatitis A virus, and Norwalk virus. For example, human fecal material from infected individuals can contain millions of shigellae per gram. Since toilet paper is unreliable in preventing fingertips from coming in contact with fecal material, fingertip washing becomes critical. Note that safe food for immune-compromised people in the U.S. has been basically defined by the government as no detectable *Salmonella* spp. or *L. monocytogenes* in one or two 25-gram samples from a lot. A level of less than 100 CFU *L. monocytogenes* per gram in ready-to-eat food, as set by Canada (46, 47), is probably realistic. Immune-compromised people must take the responsibility themselves for staying healthy.

For example, they must understand that typical foodservice salads are high-risk food items for containing *L. monocytogenes* and should be avoided. These individuals should insist on well-cooked food.

There are very few sources of data to indicate the threshold levels for hepatitis A virus and Norwalk virus. If *Shigella* spp. is controlled, hepatitis A and Norwalk virus will probably also be controlled. Assuming toilet paper is 99.9% effective, 10^6 of the 10^9 pathogens in the feces of an ill employee will get on the employee's fingertips and underneath his or her fingernails. The fingertip washing process must reduce these organisms to below 10 in order to prevent an employee from transferring this pathogen to food that will make customers ill.

Since antimicrobial chemicals used on hands only reduce pathogens about 100 to 1, the key safety strategy involves using a good hand soap or detergent; physical agitation of the fingertips with a fingernail brush; a lot of warm, flowing water; followed by a second hand wash without the brush. A greater than 10^5 reduction can be achieved using this double hand wash procedure. Threshold levels for some chemical additives are also listed in Table below. There are hundreds of additives that are accepted for food use. The addition of these compounds to food must comply with level of use defined by the Code of Federal. It is essential that all chemical additives must be measured correctly before being added to food. For instance, in foodservice, there are no guidelines for the amount monosodium glutamate (MSG) in food. Hence, MSG is commonly overused in some restaurants. The 0.5% limit of use for monosodium glutamate suggested in Table below is based on 1/3 of the levels recommended by a noted supplier of MSG

Assumed Contamination Levels for Raw Food

A series of beginning contamination levels for raw food coming into typical foodservice systems is shown in Table. It is based on threshold levels that make healthy people ill and normal contamination as listed in the literature.

These figures pertain to food in the U.S. Contamination levels in other countries may be higher or lower. When designing a safe food process, these are the levels of contamination that must be controlled and eliminated by methods of processing, preservation, and storage.

Table 4-3. Expected Per Gram Pathogen Contamination on Raw Food

Microorganisms	Meat & Poultry (CFU / g)	Fish & Shellfish (CFU / g)	Fruits & Vegetables (CFU / g)	Starches (CFU / g)
Salmonella spp., Vibrio spp., Hepatitis A, Shigella spp., E. coli, L. monocytogenes	10	10	10	10
Campylobacter jejuni	1,000*			
Clostridium botulinum	0.01	0.01	0.01	0.01
Clostridium perfringens	100	10		
Bacillus cereus	100	10	100	100
Mold toxins	Below government tolerances			
Chemicals & poisons	Below government tolerances			

* In poultry

Often there is the question for retail food operators, "What is a microbiological standard for good food quality?" Neither the USDA nor the FDA provides an answer to this question. Microbial specifications that can be used when communicating with suppliers as to the microbial quality of raw food (171).

The purpose of foodservice is to provide food that is safe and pleasurable to consume. Therefore, in addition to providing safe food by controlling and eliminating pathogens in food, the growth of spoilage

Food Safety

microorganisms must be limited and controlled as much as possible in order to provide food products of a specified quality.

Assumed Microbiological Criteria for Food Handlers and Food Contact Surfaces

In addition to microbial contamination expected in food, it is also necessary to define contamination levels for food handlers, facilities, and equipment. Based on the estimate that as many as 16 million people may get foodborne illness each year (15), for a period of 2 to 5 days. It can be estimated through calculations that 1 in 50 people working in the 500,000 food establishments in the U.S. is shedding billions of pathogens in his or her feces. If 0.001 gram of fecal material leaks through or gets around toilet paper, fingertips and underneath fingernails will become contaminated with fecal pathogens.

Hands and fingertips can also become contaminated when changing diapers, cleaning up vomit, or cleaning up after animals at home. The double hand wash method using a fingernail brush must be used to reduce fecal microorganisms to less than 10 on the fingertips and under the fingernails. The facilities and equipment will also be contaminated. If the food contact surfaces are washed and rinsed every 4 hours, and if the facilities are cleaned and sanitized adequately at the end of production, the pathogenic build-up on the equipment and facilities can be kept to a safe level. Note that *L. monocytogenes* is an environmental pathogen that arrives on food or on the people who enter the facility.

It is essential that the facility be as well maintained as possible in critical locations, so that cracks in the floors, walls, or ceilings (where *L. monocytogenes* can accumulate) are minimal. When cleaning is done regularly, *L. monocytogenes* can be reduced to an undetectable level immediately after cleaning. It is also essential that food processing areas be kept at a humidity of less than 65% to minimize mold multiplication in dry food and in the environment of the facility.

In a food production area where pasteurized, cooked, cooled food is being assembled and packaged for refrigerated meals, there must be no pathogens on food contact surfaces. This is defined as no vegetative pathogens (e.g., *Salmonella* spp., *Shigella* spp., and *C. jejuni*) in a 50-sq.-cm. swabbed area of a surface. The Standards for Number of Spoilage Microorganisms on Food Contact Surfaces are shown in Table. If these standards are maintained, industry experience has shown that foods with long shelf lives can be produced.

Food Pathogen Control Data Summary

The next step is to develop the microbiological basis for the time and temperature and pH process standards. Table below provides the database for process standards development.

Since *Y. enterocolitica* and *L. monocytogenes* both begin to multiply at 29.3°F (-1.5°C) (76), food must be kept below a temperature of 30°F (-1.1°C) if the multiplication of these pathogens is to be totally stopped. This means only having frozen food, which is impractical. Another solution is to use time with temperature from 29.3 to 127.5°F (-1.5 to 52.2°C) so that the pathogens do not multiply to an unsafe level. *Salmonella* spp. will multiply at a pH as low as 4.1. Therefore, it is essential that if food such as mayonnaise is made with raw eggs (notorious for being contaminated with *Salmonella* spp.), the pH of the product must be below 4.1. Smittle (167) determined that there is no *Salmonella* spp. growth at pH 4.1 (when acidified with acetic acid), and in fact, *Salmonella* spp. in mayonnaise is destroyed when held at 70°F (21.1°C) for 72 hours.

The data for *C. jejuni* point out that it grows very poorly over a limited range and is quite easily destroyed. Therefore, the major problem with *C. jejuni* is cross-contamination, as mentioned earlier. It can be assumed that 10,000 *Campylobacter* spp. per 50 square cm (177) will be deposited on the food contact surface by raw food such as chicken, and it must be reduced to less than 100 per 50 square cm to be tolerable.

Spores of *C. botulinum* type E are inactivated at 180°F (82.2°C). However, many foods do not reach this temperature when cooked or pasteurized. It must be assumed, then, that *C. botulinum* type E and other non-proteolytic *C. botulinum* spores will survive the cooking process. Many foods, 0.25 inch below the surface, have an oxidation reduction potential at which *C. botulinum* will grow.

For control, if the food is a probable carrier of non-proteolytic *C. botulinum*, the food must be stored below 38°F (3.3°C) in order to prevent outgrowth of the spores. *Staphylococcus aureus* begins to multiply at 43°F (6.1°C) but does not begin to produce a toxin until it reaches a temperature of 50°F (10.0°C). Since there may be recontamination of food with *S. aureus* when people make salads with their hands, if salad ingredients are pre-chilled to less than 50°F (10.0°C) and are kept below this temperature when mixed, there will be no chance of *S. aureus* toxin production. Food containing 1,000 *S. aureus* organisms per gram is not hazardous. A population of 10^6 *S. aureus* per gram of food is necessary to

Food Safety

produce a sufficient amount of toxin to cause illness. However, if the toxin is produced, it is virtually impossible to destroy the toxin when the food is cooked or reheated.

Therefore, reheating food to 165°F (73.9°C) should never be used as a critical control procedure. After food is cooked, the best method of control is to prevent the production of toxin by controlling cross-contamination and by keeping the temperature of food below 50°F (10.0°C). Some types of *B. cereus* begin to multiply below 40°F (4.4°C) (192). It is a very common contaminant of many cereal products, spices, and even fresh sprouts. This spore-forming pathogen will survive cooking. To be safe for long-term storage (greater than 10 days), cooked food must be cooled and maintained at less than 38°F (3.3°C) in order to prevent the growth of both type E *C. botulinum* and *B. cereus*.

Proteolytic strains of *C. botulinum* (types A and B) do not begin to multiply and produce a toxin until they reach a temperature of 50°F (10.0°C). Therefore, if produce such as fruits and vegetables, which today are frequently vacuum packaged and contaminated with low levels of *C. botulinum*, are kept at a temperature of less than 50°F (10.0°C), types A and B *C. botulinum* will not cause illness.

Finally, *C. perfringens* is considered to be another control organism. Actually, the highest known growth temperature for a foodborne pathogenic bacterium is that of *C. perfringens* at 126.1°F (52.3°C) and is referred to as the Phoenix phenomenon (164). Therefore, the upper temperature limit for pathogenic microorganism control is 126.1°F (52.3°C) [rounded to 130°F (54.4°C)]. Because of its rapid growth, as rapidly as once every 7.2 minutes at 105.8°F (41.0°C) (196), this pathogen dictates the heating and cooling rates for food. Food must be heated from 40 to 130°F (4.4 to 54.4°C) in less than 6 hours in order to assure no multiplication. Food must be cooled from 130 to 45°F (54.4 to 7.2°C) within 15 hours in order to prevent the outgrowth of *C. perfringens* spores during cooling (98).

FOOD SAFETY CONSIDERATIONS

Food safety considerations regarding organisms produced by techniques that change the heritable traits of an organism, such as rDNA technology, are basically of the same nature as those that might arise from other ways of altering the genome of an organism, such as conventional breeding.

These include:
- The direct consequences of the presence in foods of new gene products encoded by genes introduced during genetic modification;
- The direct consequences of altered levels of existing gene products encoded by genes introduced or modified during genetic modification;
- The indirect consequences of the effects of any new gene product(s), or of altered levels of existing gene product(s), on the metabolism of the food source organism leading to the presence of new components or altered levels of existing components;
- The consequences of mutations caused by the process of genetic modification of the food source organism, such as the interruption of coding or control sequences or the activation of latent genes, leading to the presence of new components or altered levels of existing components;
- the consequences of gene transfer to gastrointestinal microflora from ingested genetically modified organisms and/or foods or food components derived from them; and,
- the potential for adverse health effects associated with genetically modified food microorganisms.

The presence in foods of new or introduced genes *per se* was not considered by the Consultation to present a unique food safety risk since all DNA is composed of the same elements.

SAFETY ASSESSMENT

The Concept of Substantial Equivalence

Food safety is defined as providing assurance that food will not cause harm to the consumer when it is prepared and/or eaten just as to its intended use. The report of the 1990 joint FAO/WHO consultation established that the comparison of the final product with one having an acceptable standard of safety provides an important element of safety assessment. The Organization for Economic Cooperation and Development has elaborated this concept and advocated that the concept of substantial equivalence is the most practical approach to address the safety evaluation of foods or food components derived by modern biotechnology.

Substantial equivalence embodies the concept that if a new food or food component is found to be substantially equivalent to an existing food or food component, it can be treated in the same manner with respect to safety. Account should be taken of any processing that the food or food component may undergo as well as the intended use and the intake by the population. Establishment of substantial equivalence is not a safety assessment in itself, but a dynamic, analytical exercise in the assessment of the safety of a new food relative to an existing food. The comparison may be a simple task or be very lengthy depending upon the amount of available knowledge and the nature of the food or food component under consideration.

The reference characteristics for substantial equivalence comparisons need to be flexible and will change over time in accordance with the changing needs of processors and consumers and with experience. The assessment of the safety of genetically modified organisms must address both intentional and unintentional effects that may result as a consequence of the genetic modification of the food source. These effects may also arise from food sources derived from conventional breeding. Genetic modification of an organism can result in unintended effects on the phenotype of that organism, such as changes in growth or reduced tolerance to environmental stress, which are readily apparent and typically eliminated by appropriate selection procedures.

However, other unintended effects such as alterations in the concentration of key nutrients or increases in the level of natural toxicants cannot be readily detected without specific safety assessment. An assessment of substantial equivalence can be carried out at the level of the food or food component which will be used as human food. Where possible, a determination of substantial equivalence should consider comparisons as close to the species level as possible in order to allow the flexible use of many types of food products from the species in question. This will entail consideration of molecular characterization, phenotypic characteristics, key nutrients, toxicants and allergens. For certain food products, comparison at the food product level would permit a conclusion of substantial equivalence even though a comparison at the species level would consider the product to be substantially equivalent except for a defined difference. The approach is to compare the food or food component from the genetically modified organism to the range of values obtained for the traditional counterpart, taking into account the natural variation in range for the host organism and for foods and food components obtained therefrom.

The data required to demonstrate substantial equivalence may come from a variety of sources including existing databases, the scientific literature or data derived from the parental and/or other traditional strains/varieties. Regardless of the techniques used to produce new food organisms, attention must be paid to the impact of growth conditions on levels of nutrients and toxicants; for example, in the case of new plant cultivars, attention must be paid to the impact of different soils and climatic conditions. This comparative approach should lead to one of three possibilities. It may be possible to demonstrate that a genetically modified organism, or a food or food component derived from it, is substantially equivalent to a conventional counterpart already available in the food supply. If it is not possible to demonstrate substantial equivalence, it may be possible to demonstrate that the genetically modified organism or food/component derived from it is substantially equivalent to its conventional counterpart apart from certain defined differences.

Finally, it may not be possible to demonstrate substantial equivalence between the genetically modified organism or food/component derived therefrom and a conventional counterpart, either because differences are not sufficiently welldefined or because there is no appropriate counterpart with which to make a comparison. While there may be limitations to the application of the substantial equivalence approach to safety assessment, this approach provides comparable or increased assurance of the safety of food products derived from genetically modified organisms relative to food products derived by conventional methods. The Consultation recommended that safety assessment based upon the concept of substantial equivalence be applied in establishing the safety of food products derived from genetically modified organisms. Further strains/varieties may be derived from genetically modified organisms by conventional techniques, such as traditional animal or plant breeding. Where the genetically modified organisms have been determined to be acceptable as a result of the safety assessment, these further strains/varieties should be assessed on their own merits just as to practices applied for the assessment of conventionally-derived organisms.

Foods or Food Components

Characterization of the Modified Organism

It is necessary to gather information to characterize the genetically modified organism from which a food under consideration is derived. This information is needed before decisions can be taken regarding the parametres to be examined in establishing whether or not substantial equivalence exists between a new food and an existing food. Several of the strategies and guidelines which currently exist to address genetically modified food products were reviewed and the following relevant information from these might be considered:

Host

Origin; taxonomic classification; scientific name; relationship to other organisms; history of use as a food or as a food source; history of production of toxins; allergenicity; infectivity; presence of anti-nutritional factors and physiologically active substances in the host species and closely-related species; and significant nutrients associated with the host species.

Genetic Modification and Inserted DNA

Vector/gene construct; description of DNA components, including source; transformation method used; and promoter activity.

Modified Organism

Selection methods; phenotypic characteristics compared to host; regulation, level and stability of expression of introduced gene(s); copy number of new gene(s); potential for mobility of introduced gene(s); functionality of introduced gene(s); and characterization of the insert(s).

Determination of Substantial Equivalence: Characterization of the Food Product

A determination of substantial equivalence can be carried out at the level of the food source or the specific food product. This entails a consideration of the molecular characterization of the new food source; phenotypic characteristics of the new food source in comparison to an appropriate comparator already in the food supply; and the compositional analysis of the new food source or the specific food product in comparison to the comparator. The comparison may be made to the parental line/strain and/or other edible lines/strains of the same species, or it can build on a comparison of the derived food product with the analogous conventional food product.

The data required to demonstrate substantial equivalence collections, the scientific literature or specific analyses carried out on the modified food product with the conventional food product serving as a concurrent control. Substantial equivalence is established by a demonstration that the characteristics assessed for the genetically modified organism, or the specific food product derived therefrom, are equivalent to the same characteristics of the conventional comparator. The levels and variation for characteristics in the genetically modified organism must be within the natural range of variation for those characteristics considered in the comparator and be based upon an appropriate analysis of data.

Phenotypic Characteristics

For plants this would include: morphology, growth, yield, disease resistance, and other characteristics which would normally be measured by plant breeders for a given crop. For microorganisms this would include: taxonomic characterization, colonization potential, infectivity, host range, presence of plasmids, antibiotic resistance patterns, and toxigenicity. For animals this would include: morphology, growth, physiology, reproduction, health characteristics and yield.

Compositional Comparison

The compositional analysis of a genetically modified organism, or a specific food product derived therefrom, should provide sufficient information on composition to allow an effective comparison to a conventional comparator already available in the food supply, for the purpose of determining substantial equivalence. Critical components are determined by identifying key nutrients and toxicants for the food source in question. Analyzing a broader spectrum of components is in general unnecessary, but should be considered if there is an indication from other traits that there may be an unintended effect of the genetic modification. Key nutrients are those components in a particular food product which may have a substantial impact in the overall diet. These may be major constituents or minor compounds.

The determination of the key nutrients to be assessed may be influenced in part by knowledge of the function and expression product of the inserted gene. Key toxicants are those toxicologically significant compounds known to be inherently present in the species, such as those compounds whose toxic potency and level may be significant to health. The determination of the key toxicants to be assessed may be influenced, in part, by knowledge of the function and expression product of the inserted gene. In determining key nutrients and toxicants, differences among consumption patterns and practices in various cultures and societies must be recognized. The key nutrients and toxicants to be examined may differ in different regions, so they should be determined using consumption data for the target region. The more critical the nutrient or toxicant, the more attention needs to be paid to the implication of comparative differences when establishing substantial equivalence. Thus, some conclusions from the substantial equivalence determination may not be equally valid in all regions. However, this should not require a complete reassessment of safety in a new jurisdiction, but only consideration of those aspects that can be justified on health grounds, such as the impact of the specific nutrient content based upon composition and intake. In addition to analysis of key nutrients and toxicants, the extent of the analysis for unintended effects will, in part, be determined by the nature of the intended alteration and by the data from molecular and phenotypic characterization. Additional tests may be necessary if these analyses point to possible unintended effects.

Outcome of Assessment: Establishment of Substantial Equivalence

Products which are demonstrated to be substantially equivalent to an existing counterpart are regarded as being as safe as that counterpart and no further safety considerations than for the counterpart are necessary.

Food Components Except for Defined Differences

When a food product is determined to be substantially equivalent to an existing counterpart except for defined differences, it was concluded by the Consultation that further safety assessment should focus only on those defined differences. Typically the defined differences will result from the intended effect of the introduction of genetic material that encodes for one or more proteins that may or may not modify endogenous components or produce new components in the host organism. This category could also include products from genetic modification and that have been shown to produce an unintended substance(s) if that unintended substance(s) is clearly defined. The safety of introduced DNA and messenger RNA *per se* is not an issue. However, the stability of introduced genetic material and the potential for gene transfer are relevant issues in the assessment. Stability of introduced genetic material should be addressed during both the molecular characterization and the performance evaluation of the genetically modified organism in the development process. These processes minimize the likelihood of unintended effects in subsequent generations.

The Consultation considered that the majority of genetically modified products will result from the introduction of genetic material and therefore concentrated on the safety assessment of these types of products. However, the approaches described herein are equally applicable to the assessment of the safety of products which have been genetically modified by other means. The approach to assessing the safety of food products having inserted genetic material should focus on the gene product(s) and their function, including the products produced as a result of their function. The introduced genetic material will typically encode one or more proteins.

The safety assessment should concentrate on both the safety of the expressed protein(s) as well as the products produced as a result of the expressed protein(s). These products will most likely include; fats, carbohydrates or modified or new small molecule components. The safety assessment of proteins should focus on the structure, function and specificity of the protein(s) and its history of use in foods, if any. Information on these should be evaluated prior to deciding whether and what type of safety evaluation may be appropriate to assess the safety of the protein(s).

Proteins in general do not raise significant safety concerns due to the large protein component of the human diet. The typical eukaryotic cell contains tens of thousands of different proteins. Genetic polymorphism also contributes to the diversity of proteins in the diet. Generally proteins that are currently consumed or functionally similar to proteins known to be safely consumed would not raise safety concerns. Variation in proteins may also result from post-translational processing, for example glycosylation or methylation patterns of the host plant. Proteins that are not functionally similar to proteins known to be safely consumed should be assessed relative to their potential toxicity and allergenicity.

A very limited number of proteins are known to be toxic to vertebrates and those proteins, including bacterial and animal toxins, have been well characterized. Proteins that would raise a safety concern can be identified by knowing the source, amino acid sequence and function of the introduced gene(s)/protein(s). Sound scientific practice dictates that toxic proteins should not be introduced into food. If a gene is obtained from a source known to produce a mammalian protein toxin or if the introduced protein shares significant amino acid homology to a known mammalian protein toxin, an acute gavage or other *in vitro* or *in vivo* tests should be considered to provide assurance that the introduced protein is not toxic to mammals. Proteins that are transferred into food products should be assessed as to their potential allergenicity.

Structure, Function and Specificity

Generally, there will be significant knowledge on the structure, function and specificity of proteins introduced into foods through genetic modification. This information is key to determining what safety assessment is warranted as well as elucidating what products will be produced as a result of the biological activity of expressed protein(s). For example, a protein that performs the same or similar function as an endogenous protein will not likely raise a significant safety concern. Likewise, for a protein such as the insecticidal proteins from *Bacillus thuringiensis* that are active against target insects but not mammals, fish or non-target insects, information on the specificity and mode of action is important. *In vitro* studies showing binding of the insecticidal protein to gut tissue of target insects but lack of binding to mammalian tissue clearly provides important information to determine what additional data on protein safety is warranted.

Generally, the function of proteins that have been introduced into foods through genetic modification have been well characterized and these proteins are not known to exert toxic effects in vertebrates. If such well-characterized proteins do not exhibit unusual functions, further safety testing will generally be minimal. Demonstration of the lack of amino acid sequence homology to known protein toxins/allergens and their rapid proteolytic degradation under simulated mammalian digestion conditions, is appropriate to confirm the safety of these proteins as well as proteins that are not substantially similar to proteins that are known to be safely consumed. Demonstration of proteolytic digestion under both gastric and intestinal conditions supports the expectation that the protein would likely be degraded during food consumption/digestion.

The conditions for these digestions support the expected digestion of the introduced protein, including digestion by those individuals that have modified gastric conditions, such as achlorhydria, in which the gastric pH is elevated. For a protein that is not rapidly digested, additional testing may need to be considered. Certain groups of proteins or food products produced from the introduced proteins may require additional consideration, depending on the function of the protein or food product produced. Certain groups of proteins are known to exhibit antinutritional effects. Because processing may reduce or eliminate the toxic effects of these proteins, many foods that contain these toxic substances are deleterious when eaten raw but safe when properly processed. Sound scientific practice dictates that such toxic protein components should not be introduced into new food products, unless the resulting food is processed in a manner that would render the food safe. Proteins may be introduced that are enzymes and produce products such as carbohydrates, fats and oils, or small molecule components. Developments that affect carbohydrates will often be modifications of food starches, which are likely to affect the content of amylose and amylopectin as well as the branching of amylopectin.

Such modified starches are likely to be functionally and physiologically equivalent to starches commonly found in food and thus not raise any specific safety concerns However, if a food source organism is genetically modified to produce high concentrations of an indigestible carbohydrate that normally occurs at low concentrations, or to convert a normally digestible carbohydrate to an indigestible form, nutritional and physiological questions may arise that must be addressed. Some alterations in the composition or structure of fats or oils, such as an alteration of the saturation of unsaturated fatty acids, may have significant nutritional consequences or result in marked changes in digestibility. Such changes may warrant a change in the common or usual name of that product that reflects the new composition of the substance. Additionally, safety questions may arise as a result of the presence of fatty acids with chain lengths greater than C22; fatty acids with cyclic substitutions; fatty acids with functional groups not normally present in fats and oils; and fatty acids of known toxicity, such as erucic acid. The concept of substantial equivalence can be applied to assess the safety of these components of the food by comparing them to similar components present in other existing foods.

For example, a canola variety was recently developed that produced high levels of lauric acid, a fatty acid not normally found in canola. This fatty acid has a history of safe consumption as a significant component of edible tropical oils. Therefore, substantial equivalence at the component level was used to assess the safety of this product. The expected uses and patterns of consumption were also assessed in the overall safety assessment of this product. The "common or usual" name of this product was changed to reflect the compositional changes and altered uses. Genes may be introduced into organisms that encode one or more proteins that result in the production of a new or modified small molecule component in the host organism. The safety of these products should be assessed based on the knowledge of the product produced, the characteristics of

Food Safety

the product and the history of safe use of the same or similar product in other foods. Some of these products may require additional testing, including appropriate *in vitro* or *in vivo* testing, depending on the uniqueness of the product and the knowledge of its function and similarity to existing products used in food. The Consultation recognized the difficulties in performing *in vitro* and *in vivo* testing with whole foods and recommended that any additional testing be carefully designed with very specific objectives and using validated methods.

Products that are Not Substantially Equivalent

Up to the present time, and probably for the near future, there have been few, if any, examples of foods or food components produced using genetic modification which could be considered to be not substantially equivalent to existing foods or food components. Nevertheless, it is conceivable that with future developments in biotechnology, products could be developed which could be considered to have no conventional counterpart and for which substantial equivalence could not be applied. For example, there could be products derived from organisms in which there has been transfer of genomic regions which have perhaps been only partly characterised. If a food or food component is considered to be not substantially equivalent to an existing food/component, it does not necessarily mean it is unsafe and not all such products will necessarily require extensive testing. Such a food or food component should be evaluated on the basis of its composition and properties. However, a sequential approach to assess the safety and wholesomeness of these foods should be considered. This includes details of the host organism, the genetic modification and inserted DNA, as well as properties of the modified organism/product with respect to phenotype and chemical and nutritional composition. Where a modification has involved the insertion of genomic regions which have been poorly characterised it will also be important to consider the donor organism.

The results of this initial characterization and the role that the product is to play in the diet will determine whether further safety testing is needed. Although many protocols exist to test food ingredients, these methods were not designed to test the safety of complex whole foods. In particular, the use of animal feeding studies has many limitations because of the insensitivity of the test system in the detection of low level effects, problems of diet balancing and the problem of assigning adverse effects to specific major foods or food components. Despite these limitations, there are no alternatives at the present time, and if animal studies are deemed appropriate, their objectives should be clear and care taken in the experimental design. Because of these difficulties, a tailored testing programme should be devised for these products on a case-by-case basis depending on the information generated during the initial characterization.

A combination of *in vitro*, and specific *in vivo* animal models may need to be employed to further assess the safety of the product. Particular attention needs to be paid to bioavailability of new food components as well as wholesomeness. With respect to nutritional aspects, human studies may need to be carried out, especially where the new product is intended to replace a significant part of the diet. These should only be carried out when animal studies have shown that the product is not toxic. Attention should be paid to sensitive segments of the population. Furthermore, it will be important to take into account regional variations for foods which will have international distribution. Because there are presently no satisfactory animal test methods which can be employed for these types of products and other types of novel foods and food components, attention should be paid to the development of appropriate methods. Experience gained in assessing new foods which are not substantially equivalent to existing foods will serve in subsequent evaluations of similar types of products.

SPECIAL ISSUES

Allergenicity

A food allergy is an adverse reaction to an otherwise harmless food or food component that involves the body's immune system in the production of antigen-specific IgE1 to specific substances in foods. Surveys suggest that up to one-third of all adults believe that they, at one time or another, have had food allergies. Yet, true food allergy is estimated to affect less than 2% of the population. Children are at greater risk with up to 5% of infants having food allergies which are often outgrown. Allergic reactions can occur virtually to any food, though most reactions are caused by a limited number of foods.

The 1995 FAO Technical Consultation on Food Allergies concluded that the most common allergenic foods associated with IgE-mediated reactions and on a worldwide basis were fish, peanuts, soybeans, milk, eggs, crustacea, wheat, and tree nuts. These commonly allergenic foods account for over 90% of food allergies, although an extensive literature search has revealed more than 160 foods associated with sporadic allergic reactions. Gluten-containing cereals were also specifically added to the list established by that FAO Technical Consultation because of their implication in the etiology of gluten-sensitive enteropathy. Allergic reactions to foods due to antigen-specific IgE usually begin within minutes to a few hours after eating the offending food. Very sensitive persons can experience a reaction from exposure to trace quantities of the offending food. Life-threatening reactions can occur in some individuals particularly following large exposures to the offending food. An individual allergic to a specific food must avoid that food, in part through careful reading of food labels. This is because no treatment is yet available to prevent specific allergic reactions to food. Almost all food allergens are proteins, although the possibility exists that other food components may also act as haptens. While the crops from which staple foods are derived contain tens of thousands of different proteins, relatively few are allergenic.

The distribution of those proteins varies in different parts of the plant and can be influenced by environmental factors such as climate and disease stress. Conventional breeding introduces additional protein diversity into the food supply. However, variations in the protein composition of our diet brought about through conventional crop improvement practices have had little, if any, effect on the allergenic potential of our major foods. In contrast, altered dietary preferences can have significant implications for the development of food allergies. For example, allergy to peanut occurs at a significant frequency in North America and Western Europe but not in other countries where peanuts are less commonly eaten. Also, recent food introductions, such as kiwi fruit, have proven to be additional sources of food allergens.

These observations provide confidence that there are not a large number of potential allergens in the food supply, but show that new allergenic foods are sometimes introduced into the marketplace. Because of the above, a clear need exists to pay particular attention to allergenicity when assessing the safety of foods produced through modern biotechnology. The difficulties of predicting potential allergenicity of foods derived from genetically modified plants, animals, and microorganisms requires the examination of a number of parametres which are common to many food allergens. These characteristics facilitate the identification of potentially allergenic gene products, although a single criterion is insufficient to confirm allergenicity or lack thereof.

Relevant criteria include:

- Source of transferred genetic material: Particular caution must be exercised if the source of this material contains known allergens.
- Molecular weight: Most known allergens are between 10,000 and 40,000 molecular weight.
- Sequence homology: The amino acid sequence of many allergens is readily available.
- Heat and processing stability: Labile allergens in foods that are eaten cooked or undergo other processing before consumption are of less concern.
- Effect of pH and/or gastric juices: Most allergens are resistant to gastric acidity and to digestive proteases.
- Prevalence in foods: New proteins expressed in non-edible portions of plants, for example, are not of a concern in terms of food allergy.

In the assessment of gene products, the amino acid sequence should be compared against the database(s) of all known allergens, in order to screen for immunologically significant sequence similarities. Gene products from sources with no allergenic history and that lack immunologically significant sequence identity to known allergens should still be subjected to physicochemical evaluation. If a gene product is found to have the physicochemical characteristics of an allergen, caution must be used and regulatory agencies may wish to consider some appropriate action. For example, a gene product from an organism not commonly consumed could be expressed in a common food, exhibit a relevant sequence similarity to a known food allergen, and be resistant to acid and protease degradation.

The concern in that case would be that the allergenicity of the product would only be noted following exposure of a reasonable population number. Such an association between a modified food and allergic reactions would in all probability be more likely to be recognized if the food was identifiable. Comprehensive stepwise approaches to the assessment of potential allergenicity employing the principles are available. The Consultation noted that, unfortunately, reliable animal models for the assessment of the allergenicity of genetically modified foods do not presently exist, although the development of such models is to be encouraged. Clinical reagents and test subjects should be available to conduct a valid assessment of the potential of a gene product obtained from an allergenic food to be an allergen for an individual sensitive to the food which was the source of the gene product.

Food Safety

Thus, new proteins produced from genes derived from allergenic foods should be first subjected to *in vitro* assays which use sera from individuals documented as being sensitive to the food that is the source of the gene, in order to identify if allergens have been transferred. Negative or equivocal results from *in vitro* assays should be followed by using approved *in vivo* skin prick tests with sensitive test subjects.

Absence of allergens from known allergenic foods may be further confirmed by appropriately designed and approved challenge procedures in sensitive subjects. Foods that fail to elicit positive results in *in vitro* or *in vivo* tests should be treated like any other food in regards to allergenicity. Foods that are found to contain an allergen transferred from the organism which provided the DNA should not be considered for marketing approval unless they can be clearly identified in the marketplace and this identity would not be lost during distribution or processing. Labelling approaches may not be practical in all situations, and the specific problem of consumers who cannot read labels or who may not be provided with labels should be taken into account.

Gene Transfer from Genetically Modified Plants

The Consultation noted that the most relevant food safety issue concerning gene transfer is the potential consequence of the transfer of an introduced gene from material derived from a genetically modified organism to microorganisms in the gastrointestinal tract, in such a way that the gene can be successfully incorporated and expressed, and result in an impact on human or animal safety. Marker genes are inserted into genetically modified plants to facilitate identification of genetically modified cells or tissue during development. There are several categories of marker genes, including herbicide resistance genes and antibiotic resistance genes. Antibiotic resistance markers have been utilized during the transformation/selection process in the development of the vast majority of genetically modified plants. Their continued use in plants remains critical to the production of genetically modified plants. The Consultation therefore focused on these particular marker genes. With respect to the potential for gene transfer from genetically modified plants to microorganisms in the GI tract, the Consultation supported the conclusions and recommendations of the 1993 WHO Workshop entitled "Health Aspects of Marker Genes in Genetically Modified Plants". That workshop, as well as this Consultation, focused on the potential for transfer of antibiotic resistance genes since these genes are the most likely to raise safety concerns if they were to be transferred and expressed in gastrointestinal microflora. Were this to occur, it could potentially affect the therapeutic efficacy of antibiotics. In regards to genetically modified plants, the WHO Workshop concluded that "there is no recorded evidence for the transfer of genes from plants to microorganisms in the gut" and that there are no authenticated reports of such bacterial transformation in the environment of the human gastrointestinal tract.

The first of these conclusions was based on the judgement that transfer of antibiotic resistance would be unlikely to occur given the complexity of steps required for gene transfer, expression and impact on antibiotic efficacy.

In order for gene transfer to take place, the following events would need to occur:
- The plant DNA would have to be released from the plant tissue/cells and survive in the presence of the hostile environment of the GI tract, including exposure to gastric acid and nucleases;
- The recipient microorganisms would have to be competent for transformation;
- The recipient microorganisms would have to bind the DNA to be transferred;
- The DNA would have to penetrate the cell wall and translocate across the cell membrane;
- The DNA would have to survive the restriction/modification system developed by the microorganism to degrade foreign DNA; and,
- The DNA would have to be integrated into the host genome or plasmid, which requires at least 20 base pairs in a complete homologous DNA sequence for significant recombination at both ends of the foreign DNA.

The likelihood that foreign DNA would persist in a microorganism would be significantly enhanced under conditions that would exert selection pressure. Such conditions are generally considered to be restricted to antibiotic selectable markers and then only under the conditions of oral therapeutic use of the corresponding antibiotic. Only when an antibiotic resistance marker is under the control of an appropriate bacterial promoter would the antibiotic resistance gene potentially be expressed and thereby provide a selective advantage to a recipient microorganism. Antibiotic markers under plant promoters would not be expressed in a microorganism; therefore, in this situation the presence of the antibiotic would not provide a selective pressure.

The Consultation concluded, consistent with the WHO workshop, that as the possibility of horizontal gene transfer is considered to be vanishingly small, data on such gene transfer will only be needed when the nature of the marker gene is such that, if transfer were to occur, it gives rise to a health concern. In assessing any potential health concerns, the human or animal use of the antibiotic and the presence and prevalence of resistance to the same antibiotic in gastrointestinal microflora should be considered. Given that the likelihood of transfer of a gene from a genetically modified plant to a microorganism in the GI tract is remote but cannot be entirely ruled out, the Consultation recommended that FAO/WHO convene an expert consultation to address whether there are conditions or circumstances in which antibiotic marker gene(s) should not be used in genetically modified plants intended for commercial use and, if so, to define those conditions/circumstances. For example, the Consultation noted that the antibiotic vancomycin is critical in the treatment of certain bacterial diseases where multiple antibiotic resistance is prevalent, due to the lack of alternatives. The Consultation discussed issues related to assessing the safety of expressed proteins. In addition to the factors considered in that part, the specific issue related to the use of antibiotic marker genes expressed in the plant, is the potential to adversely effect the therapeutic efficacy of orally administered antibiotics.

Factors that should be considered in the assessment of potential impact on such antibiotic efficacy, include:

- The function and specificity of the expressed product;
- The digestibility of the expressed protein;
- The expression level of the expressed protein;
- The availability of any required cofactor in the gastrointestinal tract; and,
- The human or animal use of the antibiotic, taking into account those populations that consume the food product.

Gene Transfer from Genetically Modified Microorganisms

The Consultation noted that there are well-known mechanisms of transfer of genetic material between microorganisms, such as transduction and conjugation. Transformation of naked DNA into microorganisms in the GI tract has not been conclusively demonstrated. The probability of gene transfer in the GI tract has to be assessed in the light of the nature of the genetically modified organism and the characteristics of the gene construct. Possible consequences of a transfer event should be assessed based on the function and specificity of the transgene. The likelihood of maintenance of the transferred gene in a recipient microorganism increases if the gene confers to the microorganism a selective advantage. Factors that may enhance the selective advantage over other organisms or the colonization ability include: phage resistance, virulence, adherence, substrate utilization or production of bacterial antibiotics. If the transferred gene is not expected to enhance any of the survival characteristics of the recipient gastrointestinal microorganism, no further safety assessment concerning these characteristics would be required. If the function of the gene suggests that survival of the recipient organism would be enhanced, the possible health consequences need to be assessed, based on the function and specificity of the gene.

The Consultation affirmed the recommendations from the 1990 FAO/ WHO joint consultation regarding genetically modified microorganism including:
- That vectors should be modified so as to minimize the likelihood of transfer to other microbes; and,
- Selectable marker genes that encode resistance to clinically useful antibiotics should not be used in microbes intended to be present as living organisms in food.

Food components obtained from microbes that encode such antibiotic resistance marker genes should be demonstrated to be free of viable cells and genetic material that could encode resistance to antibiotics. The Consultation was not aware of any reports of transfer of genes from animal, plant or microbial origin into epithelial cells except for genes from infectious agents, such as viral DNA. However, even if such transfer were to occur, the transformed epithelial cells would not be maintained in the GI tract due to continuous replacement of these cells.

Pathogenicity of Microorganisms

The Consultation reviewed the 1990 report discussion of the issue of pathogenicity related to genetically modified microorganisms, and agreed that no new issues have arisen since that consultation. To summarize, microorganisms intended for use as food or in food processing should be derived from organisms that are known, or have been shown by appropriate tests in animals, to be free of traits that confer pathogenicity. Furthermore it was stated that assessment of viable genetically modified organisms as part of a food must also take into consideration characteristics that determine their survival, growth and colonizing potential in the GI tract, including the capability to undergo transformation, transduction and conjugation, and to exchange plasmids and phages. In this regard, a general principle was elaborated that design should be directed towards minimizing intrinsic traits in microbes that allow them to transfer genetic information to other microorganisms.

Genetically Modified Animals

The 1990 consultation reviewed the safety assessment of genetically modified animals and foods derived from them and concluded inter alia that:

- "Mammals are important indicators of their own safety, since adverse consequences of introduced genetic material will generally be reflected in the growth, development and reproductive capacity of the animal. The principle that healthy mammals only should enter the food supply is of itself a method of ensuring the safety of foods derived from animals. Primarily because some fish and invertebrates are known to produce toxins, the healthy animal principle does not provide the same degree of assurance that food derived from such animals is safe and should be used with caution in determining the need for additional safety assessment".

The OECD report on "Safety Evaluation of Foods Derived by Modern Biotechnology: Concepts and Principles", focused its attention on new foods of terrestrial origin and concluded that, "In general, foods from new strains of mammals and birds that appear to be in good health have proven to be as safe as the animal breeds from which they were derived." Subsequently OECD convened a workshop on "Aquatic Biotechnology and Food Safety" at which was discussed the notion that when an animal appears healthy, this is an indication that the animal is safe to eat. It was recognized that the apparent good health of aquatic food organisms *per se* is not a useful indicator of food safety, because many such species are known to contain either exogenously or endogenously derived compounds that are toxic to humans. Individuals of such species frequently appear to be in good health because they are resistant, in some degree, to these toxins. However, if the notion of healthy appearance is applied in this sense, and if it is applied in conjunction with other attributes to assess safety, then it can still have utility when applied to food and food components derived from aquatic animals.

In general, the OECD workshop on Aquatic Biotechnology and Food Safety considered it unlikely that techniques of modern biotechnology will increase the risk to human health if applied to aquatic organisms containing toxins. It is possible that modern breeding techniques could affect the metabolism and properties of such toxins, perhaps in ways that modify their effects. In such circumstances, however, the safety assessment of such organisms will depend on knowledge of, and on techniques to detect, toxins. Consequently, the fact that some aquatic food organisms might contain compounds toxic to humans does not reduce the value of the application of the concept of substantial equivalence.

The OECD workshop also concluded that no issue could be identified which reduced or invalidated the application of the principle of substantial equivalence to food or food components derived from modern aquatic biotechnology. However, it was recognized that in some instances there may be a lack of appropriate data from the conventional species. This lack of data could lead to difficulties when making comparisons with the new food or food component. This problem arises, in part, because there is less familiarity with most aquatic organisms in food production when compared with terrestrial food animals and plants. If animals are genetically modified to improve their resistance to bacteria and viruses that also represent a human health concern, appropriate hygiene measures should be taken to ensure that there are no food safety risks to consumers of the animal products.

Food Organisms Expressing Pharmaceuticals or Industrial Chemicals

Genetic modification has considerable potential to enable the production of pharmaceuticals or industrial chemicals in varieties of organisms, that are also used as sources of food. The Consultation recognised that, generally, the genetically modified organism would not be used as food without prior removal of the pharmaceutical or industrial chemical. The Consultation agreed that the safety assessment of pharmaceuticals and industrial chemicals, as such, was outside its remit. In situations in which the genetically modified organism or its products are used in food, the Consultation agreed that the concept of substantial equivalence, as developed elsewhere in this report, could be used for the safety assessment of the food. Some such foods could be substantially equivalent to existing foods, apart from well defined differences whilst others might be substantially equivalent to their conventional counterparts.

Food Safety

There might also be situations in which the food would not be substantially equivalent to an existing counterpart. In addition to concerns about food safety, the Consultation recognised that the genetic modification of food organisms to produce pharmaceuticals or industrial chemicals may raise ethical and control issues that were outside its remit because the issues were unrelated to food safety. The ethical issues relate to the scope for administering treatments to consumers without their knowledge. The Consultation agreed that this issue should be brought to the attention of WHO.

Databases

To facilitate the compositional comparisons necessary to establish substantial equivalence, it may be useful to use and even generate international databases containing validated data on the nutrient, allergen and, especially, toxicant composition of commonly used food organisms. If the genetically modified organism is being compared directly to its parent then the data will be developed when the modified organism and the parental organism are grown and analysed under a limited number of selected environments that are representative of the conditions under which the modified organism will be used commercially. Comparison of genetically modified plants with other commercial varieties will typically focus on data generated within these varieties grown within similar geographical regions and in which the new variety is intended to be grown commercially. Where compositional analyses of key nutrients and toxicants are required for the registration of a new plant variety, these data would be updated periodically for current commercial varieties.

The published literature on these parametres would also serve as a source for this information. It is important that the reference ranges represent reasonably current information since the ranges will probably change over time. In the case of plants, relevant information could be obtained from the international centres of the Consultative Group on International Agricultural Research which holds the worldwide mandate for the conservation and use of genetic resources. These include the Centres for Genetic Resources Conservation and Breeding Research on specific crops; *e.g.* the International Maize and Wheat Improvement Centre, the International Rice Research Institute, the International Potato Centre and the International Plant Genetic Resources Institute.

The FAO and the World Food Programme provide other sources of information on food composition. Databases on microorganisms, mainly in the form of type culture collections, are in existence but not all are suited for the purpose of establishment of substantial equivalence. The Consultation acknowledged that molecular databases are available and commonly used to identify genes/proteins of similar structure and/or function. These databases are also used to compare amino acid sequence homology of an encoded protein to known protein toxins or allergens. These databases should continue to expand as new genes/proteins are isolated and characterized. The Consultation pointed to the need to develop and expand databases with valid data on the content and ranges of nutrients, toxicants and allergens in food organisms used throughout the world.

The Application of rDNA Technology in Developing Countries

Recombinant DNA technology has broad application in developing countries and has the potential for very positive impact on their economies, which are frequently agriculturally based. In this context the view has been expressed that rDNA technology might be of greater importance for developing countries than for industrialized countries. In particular, developing countries look on rDNA technology as a means of addressing the need to produce sufficient quantities of nutritionally adequate and safe food for their growing populations.

The benefits of this technology are likely to impact directly on people at the production level as this technology is extremely easy to transfer, being "packaged in a seed". However, in order for the entire global population to fully benefit from rDNA technology, the safety assessment of food derived from genetically modified organisms requires trained manpower, up-to-date legislation and a food control system for its enforcement. This applies equally in all the countries of the world.

Food safety issues are not bound by national borders, and it is therefore important that countries that have inadequate resources for assessing rDNA technology and products derived from it, make special efforts to obtain these resources. Moreover, since globalization interconnects raw material production to processing and consumers of all regions of the world, it is imperative that proper safety assessment of foods and food components produced by genetic modification, be practised world wide.

VOLUNTARY FOOD SAFETY MANAGEMENT SYSTEM

The implementation of a comprehensive food safety management system to cover all processes conducted in a facility offers possible

Food Safety

advantages to an operator by providing a mechanism for achieving active managerial control of multiple foodborne illness risk factors associated with an entire operation. In other words, rather than the operator "fixing" only the specific items that you identify as lacking active managerial control during the inspection, the operator might choose to implement a comprehensive food safety management system to ensure continuous control over all foodborne illness risk factors of concern. Other advantages of using HACCP principles may include the following:

- Reduction in product loss
- Increased product quality
- Better control of product inventory
- Consistency in product preparation and processing
- Increased profit
- Increased employee awareness and participation in food safety
- ACTIVE, rather than PASSIVE, managerial control of risk factors.

It is recommended that prior to reviewing a voluntary food safety management system based on HACCP principles you read the FDA document entitled, *Managing Food Safety: A Manual for the Voluntary Use of HACCP Principles for Operators of Food Service and Retail Establishment.*

Validation

A voluntarily implemented food safety management system using HACCP principles needs to be "validated." Validation, for the purposes of this discussion, means to focus on scientific and technical information to determine if the system in place will effectively control the food safety hazards once implemented.

You may use observations, measurements, and evaluations taken in the establishment, as well as scientific studies and other reference materials such as the *Food Code* or other applicable regulations when validating food safety management systems. Since voluntarily implemented food safety management systems involve normal processes and not high-risk specialized processes that might otherwise *require* a HACCP plan, regulators or other food safety professionals should be able to validate a voluntary plan without assistance. This is especially true since the critical limits listed in the plan should either be the same or more stringent than those established by the *Food Code* or other applicable regulations.

Reviewing a voluntary food safety management system to determine whether the corrective actions and the monitoring, verification, and record keeping procedures are sufficient to support the system may be time consuming. Because of this, it may be helpful to seek expert advise from outside sources. Outside sources include, but are not limited to, members of academia, private food safety consultants, and other federal and state governmental officials.

The written plan for a voluntary food safety management system based on HACCP principles may be relatively simple and therefore probably will not include complex information that you might otherwise expect to see in a mandatory HACCP plan. You should be very flexible in the application of HACCP principles during your review. Generally, a written, voluntary food safety management system developed using the FDA document entitled, *Managing Food Safety: A Manual for the Voluntary Use of HACCP Principles for Operators of Food Service and Retail Establishment*, will include:

- Types of food included in the plan by category or by food preparation process
- Materials and equipment layout
- Formulations or recipes
- A flow diagram showing the preparation of the food
- Training plans for managers and food employees
- Scientific data or other information supporting the plan.

The proposed food safety management system should also detail:

- Significant food safety hazards
- Each Critical Control Point (CCP)
- Critical limits at each CCP
- Methods, frequency, and responsible personnel for monitoring
- Corrective actions to be taken if the critical limits at each CCP are not met
- Methods, frequency, and responsible personnel for verifying that monitoring is taking place and prerequisite programmes are being followed
- Records to be maintained.

As you review the identified hazards in the plan, it is recommended that you check to see that all control measures vital to food safety are somehow implemented in the operation. Due to the flexible nature of

Food Safety

voluntary food safety management systems, control measures, such as proper refrigeration or cooling, may be implemented as part of the establishment's Standard Operating Procedures and not as critical control points. Remember that the goal of voluntary food safety management systems is active managerial control of foodborne illness risk factors. How the establishment achieves this goal is clearly their choice.

As you review the critical limits associated with each CCP, be sure to verify that the critical limits are in compliance with the *Food Code* or other applicable regulation. If the critical limits are not the same or more stringent than those in the *Food Code* or other applicable regulation, it may be an oversight on the part of the operator or they could be conducting a specialized process without even knowing it. If the former is true, you may merely need to inform the operator of the applicable regulations for that food or process. If the latter is true, this Manual does not apply.

Your regulations will dictate how specialized processes and deviations from your code requirements are to be handled. In some jurisdictions, deviations from the requirements stated in the regulations are not allowed. In other jurisdictions, including those that have adopted the FDA *Food Code*, a variance and HACCP plan would be required.

In reviewing monitoring procedures at each CCP, it is recommended that the monitoring procedures include answers to the following questions:
- How will each CCP be monitored?
- What will be monitored at each CCP?
- When and how often will the monitoring take place?
- Who will be responsible for the monitoring?

You should also look to see that the monitoring intervals are adequate enough to ensure hazards are being controlled. For instance, if hot holding is designated as a CCP and the plan states that the manager will check the product temperature only once per day, the lack of frequent temperature checks may allow time for spore-forming or toxin-forming bacteria to grow to dangerous levels without any ability to take corrective action. It is clear to see how important adequate monitoring is to achieving active managerial control.

In reviewing the corrective actions for each CCP, it is recommended. As you look at the corrective actions the establishment has listed for each CCP, ask yourself if the procedure listed will result in safe food. If it will, then ask yourself if the procedure listed includes a mechanism for making sure that the problem does not happen again. If the answer is no to either of these questions, changes probably need to be made to the plan. The plan should also list who is responsible for taking corrective actions.

In reviewing verification procedures, look to see that the plan contains who is responsible for the verification and at what frequency. It is also suggested that voluntarily implemented food safety management systems be reviewed periodically to make sure all food safety hazards are still adequately controlled. Changes in menu items, equipment, or buyer specifications often require a change in the system. In this Manual, this review and subsequent change in the system is referred to as "revalidation."

Lastly, when record keeping procedures are reviewed, look to see that the procedures are clearly outlined including what is to be recorded and who is responsible for documenting the activities. It is recommended that you focus your review on helping the establishment determine whether or not they are using the easiest record keeping system for them, not on whether or not records should be kept for certain activities. If you can think of a more efficient record keeping system than what is being implemented, you may want to make a suggestion to the manager for his or her consideration. You may propose something that the establishment did not consider when it was developing the plan. The idea is that simple records, especially those that are already part of the establishment's normal operation, will most likely be maintained. However, the facility may be completely comfortable with the record keeping that is already specified in their plan.

If you see that records are not specified for certain CCPs but are for others, you may want to bring it to the manager's attention since records may be helpful in verifying that monitoring and corrective actions are conducted properly. Keep in mind that the facility has developed a voluntary food safety management system tailored to their needs and available resources. If the facility does not want to keep records, your opinion of what should be documented is irrelevant as long as active managerial control is achieved. Clearly your role as a *consultant* becomes particularly important with regard to your review of record keeping procedures.

Field Verification

The primary purpose of field verification is to determine whether the activities carried out in support of a validated food safety management system are conducted according to the written plan. In other words, "Is

Food Safety 81

the firm accurately doing what it said it does and are they operating according to the food safety management system they have in place?" By conducting a verification inspection, you can help an operator identify strengths and weaknesses in the system and offer suggestions for improvement.

Keep in mind that that there are many different types of food safety management systems. Some may control foodborne illness risk factors using only some of the principles of HACCP. Therefore, flexibility is an important component when providing guidance for voluntarily implemented food safety management systems using HACCP principles.

The Verification Process

The verification of food safety management systems involves three major activities:
- Document Review
- Record Review
- On-site Verification.

Step 1-Document Review

The review of the documents related to the food safety management system should be completed before you make on-site observations and can either be done at the office or at the establishment prior to the inspection. In order for you to gain a better understanding of the food safety management practices and procedures in place, several documents may be reviewed, including the following:
- Past inspection or verification reports
- Prerequisite programmes
- Training protocols
- The written system or plan in place.

A preliminary review of the food safety management system and associated documents may provide you with the following information:
- Problems noted during past inspections
- Type, frequency, and appropriateness of training given in support of the plan
- Types of potentially hazardous food and the food preparation processes
- Materials and layout of equipment used in the preparation and processing of the food

- Calibration procedures and frequency of any equipment involved
- Formulations or recipes for the food.

It is also recommended that before conducting the on-site verification you become familiar with the following:
- Significant food safety hazards
- Each CCP
- Critical limits for each CCP
- Method and frequency of monitoring
- Corrective actions to be taken when critical limits are not met
- In-house verification and record-keeping procedures.

Step 2-Record Review

The record review is a "spot check" to ensure that routine monitoring and in-house verification by management is occurring as specified in the plan. As you conduct the record review, ask yourself, "Do the records show that activities are being performed as specified in the plan?" The record review should take place after the document review because it will provide-
- A better understanding of the strengths and weaknesses in the food safety management system allowing you to concentrate on those areas needing strengthening
- An opportunity to become more familiar with the types of forms used in the operation before actually reviewing them.

There are at least 5 types of records or information generated to support the food safety management system that may be spot-checked:
- Prerequisite programme records (*i.e.* training logs)
- Monitoring records (*i.e.* time-temperature logs)
- Corrective action records (*i.e.* shipment rejection logs)
- Calibration records (*i.e.* logs of thermometer or pH meter calibrations)
- Evidence of verification (*i.e.* management oversight of activities related to the food safety management system).

To review the records, two approaches are suggested: Randomly select a variety of records, spot checking different time periods. Then review each record to verify that all the CCPs, associated critical limits, monitoring procedures and frequencies, corrective actions, verification and calibration activities have taken place on those days. For example: Pick one week from the previous month and identify the CCPs and critical limits for the processes used. Check to see if the monitoring was done properly and at the required frequency stated in the plan. If you note deviations from the critical limits, check to see that the appropriate corrective actions were documented. Additionally, check to make sure that the activities were verified and that the equipment used was properly calibrated.

Randomly select a few days of records, but focus only on the CCPs that appear difficult to monitor or that have shown record-keeping or compliance problems in the past. Use these records to review the associated critical limits, monitoring procedures and frequencies, corrective actions, verification and calibration activities for those days.

For example: Looking over past inspection reports, you see that hot holding has historically been a problem in this establishment. You may select one week at random from the past month and check to see if hot holding was monitored properly and at the required frequency, as stated in the plan. If deviations from the hot holding critical limit were noted, check to see that the appropriate corrective actions were documented. Additionally, check to make sure that the activities were verified and that the equipment used was properly calibrated.

It is also a good idea to include the current day's records in your review. Seeing the real-time activities of the plan will give you insight into the accuracy and consistency of the monitoring prescribed in the plan. Some questions to ask yourself as you review the records include:

- Do the recorded critical limits meet or exceed those specified in the plan?
- If deviations from critical limits are noted, do the records indicate that the appropriate corrective actions were taken?
- Do the records indicate the monitoring and verification frequencies and the individuals performing these duties?
- Do the records indicate that calibrations are being completed according to the prescribed frequency and method?

At the conclusion of the record review, determine if there are any patterns to the deviations. Multiple deviations at the same CCP can indicate that difficulties exist in controlling or monitoring that CCP. Such observations may trigger a revalidation of the system. Also, be sure to keep the group of records that you have reviewed with you so that you can continue to evaluate the critical limits, monitoring, corrective actions, etc. during the on-site verification portion of your inspection.

4

Food Spoilage

Food is considered contaminated when unwanted microorganisms are present. Most of the time the contamination is natural, but sometimes it is artificial. Natural contamination occurs when microorganisms attach themselves to foods while the foods are in their growing stages. For instance, fruits are often contaminated with yeasts because yeasts ferment the carbohydrates in fruits. Artificial contamination occurs when food is handled or processed, such as when fecal bacteria enter food through improper handling procedures.

Food spoilage is a disagreeable change or departure from the food's normal state. Such a change can be detected with the senses of smell, taste, touch, or vision. Changes occurring in food depend upon the composition of food and the microorganisms present in it and result from chemical reactions relating to the metabolic activities of microorganisms as they grow in the food.

CHARACTERISTICS OF FOOD SPOILAGE

Spoilage is characterized by any change in a food product that renders it unacceptable to the consumer from a sensory point of view. Knowledge of the microorganisms involved in spoilage and the metabolites associated with spoilage is needed to develop microbiological and chemical methods for evaluation of quality and shelf life. Such knowledge is important, *e.g.*, if a chemical spoilage index is to be developed or may be used to eliminate or prevent a particular spoilage promoting compound in foodstuffs.

Specific microorganisms are associated with food spoilage and resultant metabolic activities resulting in slime production or the production of ammonia and sulphur compounds, are responsible for undesirable flavours and odours. As part of the Gram negative rod, psychrotrophic, proteolytic group, the genus Chryseobacterium is considered to be an active food spoilage organism. Hydrolysis of proteins, fats and polysaccharides can cause changes in the texture of foods. Incomplete metabolism of the amino acids and fatty acids and fermentations of simple sugars can contribute to changes in flavour. Foods of mixed composition frequently undergo several simultaneous changes in odour, flavour and texture in the process of spoilage.

It is known that biogenic amines can cause migraines and headaches and that these amines may sometimes be present in dairy products, especially products like cheese and yoghurt. The determination of biogenic amines in foods is consequently important not only from the point of view of their toxicity, but also because they can be used as food spoilage indicators. The aim of this study was to use BIOLOG MicroPlates to identify the specific carbon sources utilized by the Chryseobacterium species, since the microbial degradation of similar carbon sources in foods can lead to the production of microbial metabolites, which could produce potential spoilage defects. An array of phenotypic tests was performed to investigate the metabolic activity of the Chryseobacterium species, as well as metabolites not included in the BIOLOG system. The ability to produce biogenic amines at different temperatures and sodium chloride concentrations by the Chryseobacterium species, was investigated by using a modified Niven medium.

Materials and methods

Strains and Growth Conditions Used

The type strains of seven Chryseobacterium spp. and two Elizabethkingia spp. were used for purposes of comparison. All the freeze-dried reference strains were reactivated in Nutrient Broth and checked for purity on Nutrient Agar at 25°C for 24-48 h. The strains were maintained on Nutrient Agar slants and stored at 4°C.

Utilisation of Carbon Sources

The reference strains were streaked on to Nutrient Agar plates and incubated at 25°C for 24 h. Gram staining, oxidase and catalase tests were performed on the reference strains as described by Hugo *et al.* to verify that the strains are Gram negative, oxidase positive and catalase negative. The strains were then inoculated on to Triple Sugar Iron Agar slants as described by Fankhauser.

The strains were then subjected to testing on BIOLOG GN2 MicroPlates just as to the manufacturer's protocol to differentiate between the different species and to identify the carbon sources utilized by the different species. The strains were classified in a Gram negative non-enteric group, inoculated on BIOLOG Universal Growth Agar and incubated at 25°C for 24 h. For identification purposes, the GN2 MicroPlates were visually read after 16 h and 24 h of incubation at 25°C.

Phenotypic Characterization of the Isolates

In order to supplement metabolites in the BIOLOG system and due to a lack of relevant phenotypical test results in literature, an extra battery of metabolic activity tests was performed to give further differentiation of the Gram negative, yellow-pigmented species at 25°C using the methods described by MacFaddin, Gerhardt et al., Barrow and Feltham and Hugo et al.

Production of Biogenic Amines

The ability of the bacteria to produce amines from amino acids was determined by using the method described by Bester et al. The cultures were streaked out on a modified Niven medium to which either histidine, tyrosine, tryptophan, arginine or lysine was added as substrate for decarboxylation. Unless otherwise indicated, incubation was at 25°C for 48 h. The effect of incubation temperature on amine production was examined by incubating the inoculated plates at 7, 15, 20, 25 or 30°C.

The effect of sodium chloride on amine production was examined by using the modified Niven medium containing each of the five amino acids separately in combination with each of 0, 1, 2, 3, 4 or 5 NaCl used to evaluate the effect of salt concentration on amine production. The plates were incubated for 48 h at 25°C.

Results and discussion

Carbon Source Utilization

All the reference strains were Gram negative rods, oxidase positive, catalase negative and produced an alkaline/acid reaction on the TSI slants, which indicated that only glucose was fermented and peptone utilized. Although the BIOLOG supplier recommended an incubation temperature of 30°C for the GN-NENT group, optimum growth for the range of reference strains tested was obtained at 25°C as the optimum growth temperature for Chryseobacterium species is 25°C. After 16 h of incubation, *C. gleum*, *C. indoltheticum* and *E. meningoseptica* produced a 100 per cent probability just as to BIOLOG identification.

After 24 h of inc ubation, *C. indologenes*, *C. scophthalmum* and *C. balustinum* produced a 100 per cent probability just as to BIOLOG identification. At 16 h to 24 h of incubation, the similarity index must be at least 0.50 to be considered acceptable. The similarity index was determined by Biolog's MicroLog computer software. Only 33 per cent of the reference strains produced a similarity index of less than 0.50 and produced no probability just as to BIOLOG identification after 16 h to 24 h of incubation. This is because *C. defluvii*, *C. joostei* and *E. miricola* are not present in the BIOLOG system's current database and therefore only the genus name was identified. Adequate and stable patterns on the MicroPlates were formed after 24 h of incubation. Some of the reactions on the MicroPlates were intermediate and classified as "borderline".

Although it is normal for certain genera to produce a light or faint purple colour reaction in some of the microplate wells, most reference strains gave clear "positive" reactions, which the researcher focused on after 24 h of incubation to identify the different carbon sources utilized by the reference strains. All the Chryseobacterium species and the two Elizabethkingia species utilized gentiobiose, a-D-glucose, D-mannose and succinic acid monomethyl ester. Seventy-eight per cent of the species utilized D-trehalose and Lasparagine. Sixty-seven per cent of the species utilized L-serine, L-threonine, uridine and glycerol. Fifty-six per cent of the species utilized dextrin, maltose, acetic acid, L-aspartic acid and glycyl-L-glutamic acid. Forty-four per cent of the species utilized a-cyclodextrin, L-alanyl glycine, inosine and thymidine. Thirty-three per cent of the species utilized Tween 40, Tween 80, D-fructose, a-ketobutyric acid, a-ketovaleric acid, L-glutamic acid, L-leucine and Lornithine. Twenty-two per cent of the species utilized glycogen, D-celliobiose, D-psicose, formic acid, propionic acid, glycyl-L-aspartic acid, L-alanine, Lphenylalanine and L-proline. Only 11 per cent of the species were able to utilize Nacetyl- D-glucosamine, D-arabinose, D-mannitol, ß-methyl-D-glucoside, citric acid, 2,3-butanediol and D,L-a-glycerol phosphate.

Instead of discussing the utilization of each carbon source individually, the carbon sources that produced clear positive reactions after 24 h of incubation, a could be assigned to chemical guilds of carbohydrates, carboxylic acids, polymers, amino acids and miscellaneous carbon sources. Forty-three of the carbon sources could differentiate between the species and these carbon sources included 13 amino acids, 12 carbohydrates, six carboxylic acids, seven miscellaneous carbon substrates and five polymers. *Chryseobacterium gleum* utilized 37 carbon sources, followed by *C. indologenes* with 33 and *E. meningoseptica* with 25. *Chryseobacterium balustinum* utilized 16 carbons sources, followed by *C. indoltheticum* and *C. scophthalmum* with 12 and *E. miricola* with 11. *Chryseobacterium joostei* and *C. defluvii* utilized only nine and eight carbon sources, respectively.

When comparing the chemical guilds, all the species utilized amino acids, except *C. defluvii*. *Chryseobacterium defluvii*, *C. joostei* and *E. miricola* were not able to utilize carboxylic acids, and *C. joostei*, *C. indoltheticum*, *C. scophthalmum*, and *E. miricola* could not utilize the polymers either.

The average of each chemical guild, utilized by the species was calculated by taking the sum of the per cent positive reactions, divided by the number of carbon sources, present in the chemical guild. The Chryseobacterium and Elizabethkingia species utilized the miscellaneous carbon substrates the most, followed by the carbohydrates, amino acids, the polymers and the carboxylic acids. Anaerobic decom position of amino acids may result in the production of obnoxious odours and is then called putrefaction. It results in foul-smelling, sulphur-containing products, such as hydrogen, methyl, and ethyl sulphides and mercaptans, plus ammonia, biogenic amines, indole, skatole, and fatty acids.

The oxidation/reduction of amino acids results in two organic acids, ammonia, and carbon dioxide, *e.g.*, glutamic acid, yields acetic acid, butyric acid, carbon dioxide, ammonia and hydrogen. Ayres *et al.*, the incomplete metabolism of amino acids can result in putrescence. The formation of ammonia from amino acids in meats and milk can result in alkalinization while the liberation of hydrogen sulphide from amino acids can result in sulphide spoilage. Examination of the tastes of the amino acids that occur in protein hydrolysates showed that bitterness is exclusively a characteristic of the hydrophobic L-amino acids.

The selection of carbon sources in GN plates is biased towards simple carbohydrates. Complex di-, tri-, or polysaccharides usually are hydrolysed to simple sugars before utilization. Monosaccharides would aerobically be oxidised to CO_2 and water. Some of the metabolic products resulting from carbohydrates include organic acids, alcohols, CO_2, hydrogen and H_2O. The microbial production of polysaccharides from various disaccharides, present in food can form unpleasant slime in and on food, causing the food to be both unpalatable and unacceptable to the consumer. Ayres et al. stated that the microbial fermentation of sugars can lead to souring and butyric spoilage defects.

The carboxylic acids are organic compounds containing oxygen and are weak acids. Many of these organic acids are oxidised by microorganisms to carbohydrates, causing the medium to become more alkaline. Aerobically the organic acids may be oxidised completely to CO_2 and H_2O. Acids may also be oxidized to other, simpler acids or to other products similar to those from sugars. The production of acids leads to sourness, but just as to Coultate, a-acids can be responsible for bitter tastes in foods. The polymers include polysaccharides and fatty acid esters of a polyoxyalkylene derivative of sorbitan. The polysaccharides can cause changes in food texture. Tween mixtures provide suitable conditions for the activation of both lipase and other esterases which can cause spoilage of the final products during storage. The hydrolysis of esters leads to the production of a carboxylic acid and alcohol.

Phenotypic Characteristics

All the Chryseobacterium species and the two Elizabethkingia species tested negative for the production of ammonia from peptone and arginine. *Chryseobacterium gleum, C. indologenes, C. joostei, C. scophthalmum* and *C. miricola* produced ammonia from urea. Ammonia is usually formed in appreciable amounts in the advanced stages of protein decomposition. The production of ammonia tends to cause an increase in pH. In addition to the formation of malodorous compounds, the release of large amounts of ammonia also contributes to the development of spoilage odours. All the Chryseobacterium species and the two Elizabethkingia species hydrolysed aesculin. This glycoside is hydrolysed to aesculetin and glucose.

Chryseobacterium indologenes and *C. joostei hydrolysed* starch, while *C. balustinum, C. defluvii* and *Elizabethkingia miricola* only hydrolyzed starch weakly. Starch is attacked by the hydrolytic action of extracellular amylases. a-Amylase rapidly liquefies starch, simultaneously attacking many 1,4-glycosidic bonds, including those in the centre of the chain. This results in the production of maltose, glucose and oligomers with three to seven glucose residues. Because of its rapid breakdown of the macromolecular structure, the viscosity of the solution and its ability to react with iodine also declines rapidly, whilst fermentable sugars appear gradually. *Chryseobacterium indologenes, C. indoltheticum* and *E. miricola* produced hydrogen sulphide. This may result from the decomposition of organic sulphur compounds, *e.g.*, cysteine and cystine, or from the reduction of inorganic sulphur compounds, *e.g.*, sulphite. Hydrogen sulphide combines with muscle pigment in meat, to give a green discolouration. Ayres *et al.*, stated that the liberation of H_2S from amino acids results in sulphide spoilage. The production of H_2S is recognized by its foul odour and is very toxic. Elizabethkingia meningoseptica was the only species to produce phenylpyruvic acid weakly. The deamination of phenylalanine leads to the production of phenylpyruvic acid and ammonia which can contribute to putrescence and alkalinization. Only C. scophthalmum was not able to produce indole from tryptophan. Organisms containing the enzyme, tryptophanase, break down tryptophan into indole, pyruvic acid, and ammonia. Indole imparts disagreeable odours associated with putrefaction. Ayres *et al.*, the formation of indole from tryptophan can lead to "unclean" flavours. *Chryseobacterium gleum* and *C. balustinum* reduced nitrate and only *C. gleum* reduced nitrite.

To Schlegel, the reduction of nitrate and nitrite can lead to the production of gaseous nitrogen dioxide and molecular nitrogen. NO_2 can cause irritating odours and is very toxic, while N_2 is odourless and non-toxic. *Chryseobacterium balustinum* and *E. miricola* produced acid from ethanol while *C. scophthalmum* produced acid weakly from ethanol. Ayres *et al.*, found that the oxidation of ethanol to acetic acid can produce a souring acetic type of spoilage in the presence of low pH foods.

Biogenic Amines

The effect of incubation temperature on the ability of Chryseobacterium species to produce spermine, cadaverine, histamine, tyramine and tryptamine on modified Niven medium. Spermine, cadaverine and tyramine were produced by all the strains at 25°C and 30°C. Only *Elizabethkingia meningoseptica* produced histamine well at 25°C. *Chryseobacterium gleum, C. indologenes, C. indoltheticum, E. meningoseptica* and *E. miricola* produced tryptamine at 25°C and C. gleum, C. defluvii, C. indologenes, C. joostei, E. meningoseptica and E. miricola produced histamine at 30°C. Temperatures at and above 25°C seemed to have a definite inhibitory effect on the production of some biogenic amines such as histamine and tryptamine.

Spermine, cadaverine and tyramine were produced by all the strains at 20°C, except for *C. defluvii* which produced cadaverine weakly at 20°C and *E. meningoseptica* produced tyramine weakly at 20°C. Spermine was produced by *C. balustinum, C. joostei* and *C. scophthalmum* at 15°C.

Food Spoilage

Cadaverine was produced by *C. balustinum*, *C. indoltheticum* and *C. joostei* at 15°C. None of the strains were able to produce histamine at 15°C. Tyramine was produced by *C. gleum*, *C. balustinum*, *C. indoltheticum*, *C. scophthalmum* and *C. joostei* at 15°C, while 67 per cent of the species produced tryptamine weakly. Only tyramine was produced at 7°C by *C. balustinum* and *C. joostei*. In a study by Bester *et al.*, flavobacterial strains of cluster 1A, were able to produce tyramine at 7°C. No species were however able to produce spermine, cadaverine, histamine and tryptamine at 7°C.

Temperatures at and below 15°C seemed to have a definite inhibitory effect on the production of spermine, cadaverine, histamine, tyramine and tryptamine. However, the ability of some Chryseobacterium spp. to produce biogenic amines, at low temperatures, *i.e.*, 15°C and 7°C, is of great significance to the dairy and other food industries.

Refrigerated products could cause amine poisoning if they were contaminated with amineproducing bacteria, *e.g.*, Chryseobacterium spp. It must be taken note of that most amines, including histamine, are heatstable and only partially destroyed in 3 h at 102°C or 90 min at 116°C. These findings stress the importance of the hygienic production and handling of food products and maintenance of the cold chain throughout production and distribution. The production of spermine, cadaverine, histamine, tyramine and tryptamine on modified Niven medium to which various amounts of sodium chloride were added. Spermine, cadaverine, histamine and tryptamine were not produced by the strains tested in the presence of 4 per cent and 5 per cent NaCl.

It would seem that an increase in the salt concentration resulted in fewer strains being able to produce the biogenic amines. Exceptions were *C. gleum*, *C. indoltheticum* and *E. miricola* that produced tyramine weakly in the presence of 4 per cent NaCl. Only *C. gleum*, *C. indologenes*, *C. joostei* and *E. meningoseptica* produced tyramine in the presence of 3 per cent NaCl. Spermine was produced by *C. gleum*, *C. indoltheticum*, *C. joostei*, *E. meningoseptica* and *E. miricola* in the presence of 2 per cent *NaCl*. *Elizabethkingia meningoseptica* produced cadaverine in the presence of 2 per cent NaCl, while only *C. joostei* was able to produce histamine weakly in the presence of 2 per cent NaCl. Tyramine was produced by *C. joostei* and *E. meningoseptica* in the presence of 2 per cent NaCl. Spermine, cadaverine and tyramine were produced by all the strains in the presence of 1 per cent NaCl, with the exception of *C. defluvii* that produced cadaverine and tyramine weakly. It would seem that the decarboxylation of histidine and tryptophan were very sensitive to NaCl concentrations, because only C. joostei was able to produce histamine in the presence of a 1 per cent NaCl, while *C. gleum*, *C. defluvii* and *C. indologenes* were able to produce tryptamine at this concentration. These results show that salt concentrations in excess of 4 per cent would be needed to prevent amine production of Chryseobacterium species in food products. This is in contrast to a study by Bester *et al.*, who showed that 7 per cent of the flavobacterial strains of cluster 1A tested by them were able to produce tyramine even in the presence of 5 per cent NaCl.

TYPES OF SPOILAGE

Various physical, chemical, and biological factors play contributing roles in spoilage. For instance, microorganisms that break down fats grow in sweet butter (unsalted butter) and cause a type of spoilage called rancidity.

Certain types of fungi and bacteria fall into this category. Species of the Gram-negative bacterial rod Pseudomonas are major causes of rancidity. The microorganisms break down the fats in butter to produce glycerol and acids, both of which are responsible for the smell and taste of rancid butter. Another example occurs in meat, which is primarily protein. Bacteria able to digest protein (proteolytic bacteria) break down the protein in meat and release odoriferous products such as putrescine and cadaverine. Chemical products such as these result from the incomplete utilization of the amino acids in the protein.

Food spoilage can also result in a sour taste. If milk is kept too long, for example, it will sour. In this case, bacteria that have survived pasteurization grow in the milk and produce acid from the carbohydrate lactose in it. The spoilage will occur more rapidly if the milk is held at room temperature than if refrigerated. The sour taste is due to the presence of lactic acid, acetic acid, butyric acid, and other food acids.

Sources of microorganisms

The general sources of food spoilage microorganisms are the air, soil, sewage, and animal wastes. Microorganisms clinging to foods grown in the ground are potential spoilers of the food. Meats and fish products are contaminated by bacteria from the animal's internal organs, skin, and feet. Meat is rapidly contaminated when it is ground for hamburger or sausage because the bacteria normally present on the outside of the meat move into the chopped meat where there are many air pockets and a rich supply of moisture. Fish tissues are contaminated more readily than meat because they are of a looser consistency and are easily penetrated. Canned foods are sterilized before being placed on the grocery shelf, but if the sterilization has been unsuccessful, contamination or food spoilage may occur. Swollen cans usually contain gas produced by members of the genus Clostridium. Sour spoilage without gas is commonly due to members of the genus Bacillus.

This type of spoilage is called flat-sour spoilage. Lactobacilli are responsible for acid spoilage when they break down the carbohydrates in foods and produce detectable amounts of acid. Among the important criteria determining the type of spoilage are the nature of the food preserved, the length of time before it is consumed, and the handling methods needed to process the foods. Various criteria determine which preservation methods are used.

MICROBIOLOGICAL ASPECTS OF FOODS

Water activity, a_W, is a physical property that has a direct implication for microbiological safety of food. Water activity also influences the storage stability stability of foods as some deteriorative processes in foods are mediated by water. Storage life of dry foods such as biscuits is generally longer then of moist foods such as meat at the same temperature. In this connection freezing of foods is equivalent to drying. The water is removed from the food matrix although it is still in the food as ice. There is strong association between the physical property, a_w, and the chemical and microbological properties of food.

Food Poisoning, Food-borne Infections

Food poisoning is the result of ingesting a pre-formed toxin in food. These toxins may result in vomiting (e.g., *S. aureus* or *Bacillus cereus* enterotoxins), or other systemic effects, e.g., botulinal neurotoxin paralysis of the nerve-muscle junction. Food-borne infections result from ingesting an organism capable of surviving the acidic environment of the stomach and growing in the intestinal tract, e.g. *Salmonella* spp. Gastrointestinal symptoms, e.g., diarrhoea, result from toxic metabolites produced in the gut.

Spoilage

Microbial spoilage of foods results from changes in the food composition, and/or appearance or structure as a result of the growth and metabolism of microorganisms. Commonly the evolution of obnoxious odours is the cause for rejection of foods, e.g., fresh meats, although the appearance of mould colonies on semi-dry foods, e.g., bread, cheeses, is also common.

A wide range of organisms can be responsible for spoilage, and therefore a wide range of changes in foods may be regarded as spoilage. Certain .controlled spoilage by micro-organisms is used to produce a different food from the starting ingredients, e.g., yoghurt or cheese from milk, fermented sausages from raw meats, sauerkraut from shredded cabbage.

Solute Effects on Microbial Growth and/or Death

Cell Membrane Phenomena

The cell membrane is semi-permeable, or rather selectively permeable. Thus glycerol penetrates the membrane readily, glucose penetrates poorly, sucrose very poorly, and NaCl is almost non-penetrating. When an organism is grown or exposed to low a_w conditions, the cells may accumulate from the environment or synthesize compatible solutes, e.g., glutamine, proline, betaine in bacteria, trehalose in yeasts.

These internal solutes interfere little with the metabolism of the cell, although metabolic energy must be diverted for synthesis, but increase resistance to low external a_w conditions, and also increases resistance to other injurious treatments, e.g., heat. This effect differs with different external solutes, e.g., *S. aureus* synthesizes compatible solutes at high NaCl levels, but not in the presence of sugar. If the partially dehydrated cell is exposed to a high temperature, then the micro-organism displays a greater thermal resistance than when grown at a higher a_W. Proteins and other essential cellular components are more resistant to thermal damage in the partially dehydrated state.

Water activity plays an important role in the heat resistance of microbes. Death curves are not always linear and interpolation of D-values (and z-values) into application of thermal processes may not always be safe. Similarly, the ratio of effects of the different solutes on D-values differs for each organism.

Thus D-values in low a_w solutions or foods must always take into account the actual solute controlling the a_W, and if necessary be determined in that solute. Interactions with other physico-chemical parameters Since extra energy is required to combat the inimical effects of low a_W, any other conditions also requiring expenditure of extra energy, e.g., low pH, presence of preservatives, will result in an additive or synergistic effect in limiting microbial growth.

Thus even moderate reductions in a_w in combination with low levels of preservative or pH values, can be sufficient to inhibit growth. One good example is that of inhibition of *Clostridium botulinum*. Under ideal conditions 10% NaCl is required to inhibit the proteolytic species; at the pH values typical of meats (ca. pH 5.4 . 5.8) and in the presence of ca 100ppm nitrite, only 3.5% NaCl is required to produce botulinal-stable cured meats.

MICROBIAL METABOLITES

Biological as well as fabricated food structures will possess receptors to which microorganisms can absorb. The resulting colonization of such structures may occur in a stratified way, leading to relatively high local concentrations of microbial metabolites. The metabolites formed by a given spoilage association will once again depend on the prevailing intrinsic, extrinsic and implicit conditions.

These include the limiting factors influencing:
- The type of spoilage, determined by the relative amounts of metabolites formed; and
- The rate at which these metabolites are produced during storage and distribution of the food.

The latter is mostly expressed as the time to spoilage, as detected by sensory evaluations—odour, colour, structure and taste. The microbial metabolites depend not only on the storage conditions but also on other environmental factors such as aeration, glucose and lactate availability, and pH.

Carbohydrates

Carbohydrates, if available, usually are preferred by microorganisms to other energy-yielding foods. The carbohydrates are divided into monosaccharides, disaccharides, and polysaccharides. The monosaccharides are polyhydroxy aldehydes or polyhydroxy ketones. For utilisation, bacteria first need to break down complex carbohydrates such as starch into their constituent monosaccharides. The random splitting of glycosidic bonds results in softening and liquefaction. Several bacteria possess an extracellular enzyme, diastase or amylase, which hydrolyses the starch. The starch is then converted either directly to glucose or via intermediates such as maltose. Although flavobacteria do not degrade lignin and cellulose, it is possible that these organisms are involved in the breakdown of various proteins and carbohydrates. Glucose is the main carbohydrate used as a carbon and energy source.

The breakdown of this monosaccharide can proceed by several pathways. In aerobic respiration the glucose metabolite, pyruvate is converted into carbon dioxide and water by means of the tricarboxylic acid cycle, Krebs cycle, or citric acid cycle. To enter the system, the pyruvate is converted to acetate activated by coenzyme A. Only the aerobic and some facultatively anaerobic microorganisms possess an intact TCA cycle. The pyruvic acid can be decarboxylated to form acetaldehyde and CO_2. The acetaldehyde can remain or be reduced to ethyl alcohol, oxidized to acetic acid, or condensed to form acetoin or acetylmethylcarbinol. The AMC can be oxidized to diacetyl, which has a butter flavour, or reduced to 2,3-butanediol. Pyruvate can be aminated to form alanine. Boers *et al.* observed that the glucose concentration had decreased to a low level at the first signs of spoilage. It has been concluded also that glucose limitations caused a switch from a saccharolytic to an amino acid degrading metabolism in at least some bacterial species.

Foods with high levels of carbohydrates are preferentially colonized by glycolytic organisms and tend to ferment rather than putrefy. This leads to the production of acids and is accompanied by a reduction in pH. The lactate occurring in flesh foods due to post mortem glycolysis can often be differentiated by its optical rotation from lactic acid formed by microorganisms; this increases its reliability as an index of spoilage. However, in some instances lactic acid may be dissimilated and acetic acid may be a better indicator of microbial colonization and metabolism.

Fats

The principle lipids in foods are fats. Fats are esters of glycerol and fatty acids and are called glycerides, in the ratio of one molecule of glycerol to three molecules of fatty acids. A pure fat is not attacked by microorganisms, since there must be a nutrient-containing aqueous phase in which the organism can grow. Lipase, an enzyme that hydrolyses fats to free fatty acids and glycerol, is present in many kinds of foods.

Because milk contains an appreciable amount of this enzyme, milk fat often undergoes lipase-catalyzed hydrolysis with the production of free fatty acids, diglycerides, monoglycerides, and in extreme cases, free glycerol. Short-chain water-soluble fatty acids cause obnoxious rancid flavours in milk. Lipolysis in foods followed by ß-oxidation produce ketones, which always result in off-flavours. The oxidative deterioration of fats involves the reaction of unsaturated fatty acids with oxygen to yield hydroperoxides. The hydroperoxides are not flavour compounds, but readily decompose to carbonyl compounds resulting in off-flavours or - odours. The carbonyl compounds are mixtures of saturated and unsaturated aldehydes and produces ketones.

Proteins

Microorganisms, through their proteolytic enzymes, break down protein into simpler substances. The breakdown usually follows the following pattern:

Protein →Peptones →Polypeptides →Peptides →Amino acids →Ammonia → Elemental nitrogen.

Proteinases catalyze the hydrolysis of proteins to peptides, which may impart a bitter taste to foods. Peptidases catalyse the hydrolysis of polypeptides to simpler peptides and finally to amino acids. The latter impart flavours, desirable or undesirable, to some foods; *e.g.*, amino acids contribute to the flavour of ripened cheeses.

The products that are formed depend upon the type of microorganism; the types of amino acids; temperature; the amount of available oxygen; and the types of inhibitors that might be present. Decomposition of protein by aerobic organisms is called decay. Proteins containing amino acids with sulphur, such as cystine and methionine, can be broken down with no unpleasant odour because the end products are completely oxidized and stabilized. Sulphur compounds, however, are often associated with "putrid" odours. The metabolites produced by microorganisms in proteinaceous foods such as meat include ammonia, ethanol, lactate, acetate, indole and acetoin, with smaller quantities of higher fatty acids, amines and ethyl esters of the lower fatty acids, sulphides, hydrogen sulphide and mercaptans. Most of the esters, amines, ammonia and sulphur compounds are produced from amino acids.

There is no significant degradation of protein proper until spoilage has progressed to obvious deterioration. Owing to production of amines and ammonia, the pH of proteinaceous foods tends to rise as spoilage progresses. An increase in the pH of a protein food indicates protein degradation, just as a decrease in pH results from the fermentation of carbohydrates.

Volatile compounds

In fresh products, such as fruit, vegetables and milk, flavour components are very abundant. Chang stated that while the odour of some foods may be accounted for by single key compounds, most food odours are the result of complex mixtures. Dainty et al. stated that the variability of individual chemicals found in the aroma of spoiled samples was not significant. A better understanding of the complexity can be gained if the volatile compounds are grouped into classes: sulphur compounds, ketones, esters, aromatic hydrocarbons, aliphatic hydrocarbons, aldehydes and alcohols, but just as to Fedele et al. found that not all of these compounds have significant effects on the overall odours. Examination of volatile compound profiles indicated there were at least 3 requirements for development of putrid odours.

These requirements are that:
- The total volatile compound peak area must be appreciably high,
- With exception of aliphatic hydrocarbons, the sulphur compounds must be the major constituents of the profile, and
- Large quantities of other classes, if present, may modify the effects of the sulphur compounds.

The determination of total "volatile bases", which include ammonia, trimethylamine and other compounds, correlates well with organoleptic judgement in a number of species of fish. Stutz et al. found that the concentration of four of the volatile compounds, acetone, methyl ethyl ketone, dimethyl sulphide and dimethyl disulphide increased continuously during storage of minced meat stored aerobically at 5, 10, or 20°C. Hydrogen sulphide and ammonia are formed as a result of the conversion of cysteine to pyruvate by the enzyme cysteine desulphydrase. Acetoin is the major volatile compound produced on raw and cooked meats in O_2-containing atmospheres. Overton and Manura milk samples were found to contain numerous straight and branched chain hydrocarbons, aldehydes, alcohols, ketones, fatty acids, esters, phenolic compounds and lactones.

Biogenic amines

Biogenic amines are basic nitrogenous compounds formed mainly by decarboxylation of amino acids or by amination and transamination of aldehydes and ketones. Biogenic amines in food and beverages are formed by the enzymes of raw material or are generated by microbial decarboxylation of amino acids, but it has been found that some of the aliphatic amines can be formed *in vivo* by amination from corresponding aldehydes. Koessler *et al.* proposed that biogenic amine formation is a protective mechanism for bacteria against acidic environments. The production of amines requires the availability of free amino acids and appropriate status of environmental factors such as pH and temperature.

The pre-requisites for biogenic amine formation by microorganisms are:

- Availability of free amino acids, but not always leading to amine production;
- Presence of decarboxy lase-positive microorganisms; and
- Conditions that allow bacterial growth, decarboxylase synthesis and decarboxylase activity.

Biogenic amines are present in a wide range of food products including fish products, meat products, dairy products, wine, beer, vegetables, fruits, nuts and chocolate. Virtually all foods that contain proteins or free amino acids and are subject to conditions enabling microbial or biochemical activity, are conducive to the production of biogenic amines. The total amount of the different amines formed strongly depends on the nature of the food and the microorganisms present. Different biogenic amines have been detected in fish such as mackerel, herring, tuna, and sardines.

Other amines, such as trimethylamine and dimethylamine are present in fish and fish products at levels depending on the fish freshness. Bacterial-produced histamine has also been found in dairy products and vegetables. Amines are also important because of their role in causing spoilage of dairy products by producing typical off-flavours and putrid odours. Putrescine, cadaverine, histamine, tyramine, spermine and spermidine were found to be present in minced pork, beef and poultry stored at chill temperatures. Histamine has been recognized as the causative agent of scombroid poisoning as well as nausea, vomiting, gastrointestinal distress and headache, whereas tyramine has been related to food-induced migraines and hypertensive crisis in patients under antidepres sive treatment with monoamine oxidase inhibitor drugs. Secondary amines such as putrescine and cadaverine can react with nitrite to form heterocyclic carcinogenic nitrosamines, nitrosopyrolidine and nitrosopiperidine.

The levels reported for histamine and its potentiators in food would not be expected to pose any problem if normal amounts were consumed.

Sandler *et al.* reported that 3 mg of phenylethylamine causes migraine headaches in susceptible individuals, while 6 mg total tyramine intake was reported to be a dangerous dose for patients receiving monoamine oxidase inhibitors.

The level of 1,000 mg kg-1 is considered dangerous for health. This level is calculated on the basis of foodborne histamine intoxications related to amine concentration in food. The European Community has recently proposed that the average content of histamine should not exceed 10-20 mg/100 gm of fish.

GROWTH OF FOOD SPOILAGE BACTERIA

Temperature

One of the most crucial factors affecting microbial growth in food is temperature. Growth is restricted to those temperatures at which an organism's cellular enzymes and membranes can function. As the temperature rises, chemical and enzymatic reactions in the cell proceed at more rapid rates, and growth becomes faster. However, particular proteins may be irreversibly damaged. Thus, as the temperature is increased within a given range, growth and metabolic functions increase up to a point where inactivation reactions set in.

Every food spoilage bacteria has cardinal temperatures, namely, a minimum temperature below which growth no longer occurs, an optimum temperature at which growth is most rapid, and a maximum temperature above which growth is not possible. Since the first observation of bacterial growth at 0°C, many terms were used for these organisms. The term "psychrotroph" was introduced by Eddy to replace the misnomer "psychrophilic". The latter term indicates organisms that have a preference for growing at low temperatures, while psychrotrophs should rather be regarded as cold tolerant being able to grow at 7°C or less but having optimum temperatures of 25 to 35°C. In 1976 the International Dairy Federation adopted the following definition: A psychrotroph is a microorganism which can grow at 7°C or less, irrespective of its optimum growth temperature.

Many psychrotrophic bacteria, when present in large numbers, can cause a variety of off-flavours as well as physical defects in foods. Raw foods held under refrigeration prior to processing, as well as non-sterile heat processed foods that rely on refrigeration for shelf life, are subject

Food Spoilage

to quality loss and possible spoilage by psychrotrophic bacteria. Although psychrotrophic bacteria will not grow in frozen foods, they can grow and cause spoilage if the food is allowed to thaw partially, and is subsequently held at too high a temperature. Studies have revealed that the most common bacteria isolated on dairy equipment surfaces are Gram negative psychrotrophs which are responsible for growth and spoilage in milk at refrigeration temperatures. Jooste et al. investigated the role of flavobacteria as causative agents of the putrid butter defect and found that the optimum growth temperature for the six Flavobacterium strains from butter tested, was 25°C and that these strains were capable of multiplication in cream both at 6°C and 25°C.

pH

pH is one of the main factors affecting the growth and survival of microorganisms in culture media and in foods. All microorganisms have a pH range in which they can grow and an optimum pH at which they grow best. Bacteria generally have a minimum pH for growth of around 4.0 – 4.5 and an optimum pH between 6.8 and 7.2, and pH maxima between 8.0 and 9.0. Organisms that thrive at low pH values are called acidophiles. Organisms that have very high pH optima for growth, are known as alkaliphiles, which can produce hydrolytic enzymes, such as proteases and lipases. The pH minimum for an organism is determined by the temperature of the environment, the nutrients that are available, the water activity and the presence or absence of inhibitors.

Despite the pH requirements of a particular organism for growth, the optimal growth pH represents the pH of the extracellular environment only, the intracellular pH must remain near neutrality in order to prevent destruction of acid-or alkali-labile macromolecules in the cell. For the majority of microorganisms, whose pH optimum for growth is between 6 and 8, the cytoplasm remains neutral or very nearly so. When the microbial cell is subjected to extreme pH values, cell membranes become damaged. The pH minimum for an organism depends on the type of acid present. Generally, the minimum is higher if any organic acid is responsible for the environmental pH rather than an inorganic acid. Foods are quite variable in terms of their pH values.

Most are acidic, ranging from the very acidic to almost neutral in reaction. pH changes in foods due to the activity of microorganisms are common. Meat becomes more alkaline when spoilage is caused by Gram negative rods such as Pseudomonas spp. The organism uses amino acids as its carbon source which leads to the production of ammonia, making the cell environment more alkaline. Shimomura *et al.* found that the pH range of *Chryseobacterium shigense* for growth was 5-8. Park *et al.* found the pH range for growth of *Chryseobacterium soldanellicola* is pH 5-7 and that for optimal growth is pH 5. The pH range for growth of *Chryseobacterium taeanense* is pH 5-9 and that for optimal growth is pH 5.

Water activity and Sodium chloride

Water availability not only depends on the water content of an environment, that is, how moist or dry a solid microbial habitat may be, but is also a function of the concentration of solutes such as salts, sugars, or other substances that are dissolved in water. This is because dissolved substances have an affinity for water, which makes the water associated with solutes unavailable to organisms. Water availability is generally expressed in physical terms such as water activity. The water content of a food may bear little relationship to its water activity.

Foods may have a low salt content but a low water activity. Each specific organism has its own range of water activity in which it will grow. Most organisms have an optimum approaching 1.0, where the water activity is high but, where there is also sufficient dissolved nutrients to support rapid growth. An added complication is the reaction that some organisms show towards sodium chloride. Halophiles are organisms that require sodium ions in order to grow. Moderate halophiles are organisms that require NaCl but will grow only at moderate concentrations, *i.e.*, between 1 and 10 per cent. Sodium ions are believed to be involved with the transport mechanisms associated with the cell membrane and the uptake of materials from the environment.

Extreme halophiles are organisms that will only grow at high sodium chloride concentrations and generally require 15-30 per cent NaCl, depending on the species, for optimum growth. Halotolerant organisms can tolerate some reduction in the aw of their environment, but generally grow best in the absence of the added solute. In a study by Jooste *et al.*, the Flavobacterium strains tested in NaCl Broth were able to grow in 1 per cent NaCl, but not in 4 per cent NaCl Broth. Mudarris *et al.* found that *Chryseobacterium scophthalmum* was able to grow in the presence of 0 to 4 per cent NaCl, but not in the presence of 5 per cent NaCl. Jooste and Hugo no growth of *Chryseobacterium indologenes* and *Chryseobacterium meningosepticum* occurred in Nutrient Broth with the addition of 6 per cent NaCl. De Beer *et al.* found that *Chryseobacterium gleum* and *Chryseobacterium indologenes* exhibited growth in the presence of 3 per cent NaCl. Park *et al.* found that *Chryseobacterium soldanellicola* exhibited growth in the presence of 0 to 4 per cent NaCl within 14 days and *C. taenense* exhibited growth in the presence of 0 to 6 per cent NaCl within 14 days.

THE GENUS CHRYSEOBACTERIUM

Ecology

Flavobacteria are widely distributed in soil and aquatic environments, raw meat, and milk, and have been found in human clinical material. CDC Group IIb strains have been found to be the most common of the Flavobacterium isolates from clinical specimens and the hospital environment in the UK. The clinical ro le of Flavobacterium, including Group IIb, has been reviewed by Von Graevenitz and several reports have drawn attention specifically to the role of Group IIb in a case of meningitis and in various cases of bacteraemia. In a study of yellow pigmented Gram negative bacteria from environmental sources by Hayes, phenon 1 was found to be the largest single cluster.

Representative strains of this phenon which were subsequently examined by Owen and Holmes were found to resemble Group IIb strains. In a study by Jooste *et al.*, it was found that their cluster 1A comprised 43 per cent of the total isolates. This cluster was regarded as Group IIb-like organisms, since it contained the reference strains NCTC 10795 and strain M15/1 of Hayes phenon 1. It would appear, therefore, that this group of bacteria is generally the most prevalent flavobacterial taxon in both clinical and non-clinical environments.

Taxonomy of Chryseobacterium

The genus Chryseobacterium was proposed in the mid nineties by Vandamme *et al.* for the species in Group A of Holmes which included Flavobacterium CDC Group IIb strains. Flavobacterium species that have since been renamed include: *Chryseobacterium indologenes*, *Chryseobacterium gleum*, *Chryseobacterium indoltheticum*, *Chryseobacterium balustinum*, *Chryseobacterium meningosepticum* with *C. gleum* as type species as both its genotypic and phenotypic characteristics have been studied in detail. *Chryseobacterium scophthalmum* was also included in this genus in 1994. Since the publication of Bernardet *et al.*, new species have been validated.

Chryseobacterium defluvii, isolated from sewage water and *C. miricola*, isolated from condensation water in a Russian space station were also introduced to the study.

Studies by Hugo and Hugo et al., have shown that yellow pigmented flavobacterial strains from raw milk were actually members of the genus Chryseobacterium.

These studies also led to the description of a new species, *Chryseobacterium joostei*, isolated from raw milk. The latest validated species of this genus are *C. formosense*, *C. daecheongense*, *C. taichungense*, *C. shigense*, *C. vrystaatense*, *C. soldanellicola* and *C. taeanense*.

The species validated in 2005 and 2006 were not included in this study. One strain, "*C. proteolyticum*" has been described by Yamaguchi and Yokoe but has not been validly published. Studies by Kim et al. proposed that *C. meningosepticum* and *C. miricola* should be transferred to a new genus, Elizabethkingia. Of the currently validated species of Chryseobacterium, only *C. balustinum*, *C. gleum*, *C. indologenes* and *C. joostei* are often associated with food.

Description of Chryseobacterium

Cells are Gram negative, non-motile, non-spore-forming rods with parallel sides and rounded ends; typically the cells are 0.5 ìm wide and 1 to 3 ìm long. Intracellular granules of poly-ß-hydroxybutyrate are absent. All Chryseobacterium species are aerobic chemoorganotrophs; their metabolism is strictly respiratory, not fermentative, except for *C. scophthalmum*, which also exhibits a fermentative metabolism.

However, some such as *C. gleum*, *C. indologenes* and other CDC Group IIb strains can grow anaerobically in the presence of nitrate by using nitrate as a terminal electron acceptor and reducing it to nitrogen. *Chryseobacterium indologenes* strains can also grow under anaerobic conditions in the presence of fumarate. All strains grow at 30°C; most strains grow at 37°C. Growth on solid media is typically pigmented but non-pigmented strains occur.

Colonies are translucent, circular, convex or low convex, smooth, and shiny, with entire edges. The strains are positive for catalase, oxidase and phosphatase activities. Several carbohydrates, including glycerol and trehalose, are oxidized. Strong proteolytic activity occurs. Esculin is hydrolyzed, but agar is not digested. These organisms are resistant to a wide range of antimicrobial agents. Branched-chain fatty acids are

Food Spoilage

predominant. Sphingophospholipids are absent. Menaquinone 6 is the only respiratory quinone.

Homospermidine and 2-hydroxyputrescine are the major polyamines in C. indologenes, whereas putrescine and agmatine are minor components. Most Chryseobacterium species exhibit a rather high tolerance to sodium chloride, except C. balustinum. All Chryseobacterium species are able to grow on marine agar, although only members of two species were actually isolated from marine environments. The DNA base composition ranges from 33 to 38 mol per cent guanine plus cytosine. Chryseobacterium are widely distributed in soil, water, and clinical sources.

Contamination of foods with *Chryseobacterium*

Jooste et al. were the first to isolate Flavobacterium CDC Group IIb strains from milk and butter. In a subsequent study, it was suggested that these Flavobacterium species caused putrefaction in salted butter by growing in cream prior to churning. In another study in which Flavobacterium CDC Group IIb and C.

Balustinum strains were isolated, Jooste and Britz found that the practical importance of dairy flavobacteria lies as much in their psychrotrophic growth and consequent proteinase production in refrigerated milk as in their contamination of milk via poorly sanitized pipelines and equipment. A study by Welthagen and Jooste indicated that CDC Group IIb isolates comprised the largest part of pigmented bacteria from raw milk.

In subsequent investigations a large group of the CDC Group IIb milk isolates evaluated in the studies were identified as *C. indologenes* and one isolate as *C. gleum*. Among the remaining milk isolates, two new genomic groups were identified.

In a study by Venter et al., a metalloprotease from a strain of *C. indologenes* was purified and characterized. This protease was very heat-stable and its affinity for casein may play a role in the spoilage of milk and milk products.

Several Chryseobacterium species are associated with spoilage of dairy products during cold storage and include *C. balustinum*, *C. gleum* and *C. joostei*. *Chryseobacterium balustinum* had initially been isolated from the scales of freshly caught halibut in the Pacific Ocean. Since this organism produced a yellowish slime on the skin, it was considered a fish spoilage agent rather than a pathogen. Although, recently isolated again from the skin and muscle of wild and farmed freshwater fish, *C. balustinum* was not regarded as an important contributor of the spoilage of the fish because of its low incidence.

The five *C. balustinum* strains, isolated in the latter study and identified following a rather extensive phenotypic characterization, were not found in freshly caught fish, but in fish stored more than three days in melting ice.

Gennari and Cozzolino isolated 39 strains of flavobacteria from the skin and gills of fresh and ice-stored Mediterranean sardines. Analysis of their phenotypic traits, however, could not place the isolates in any known species of the flavobacteria, but the authors found that four strains had characteristics resembling those of Holmes' group A. Most of the species in this group are now known as Chryseobacterium.

Flavobacteria have been frequently isolated from meat and poultry products, although they have seldom been accurately identified. When Hayes divided a large collection of flavobacteria and related Gram negative yellow pigmented rods into nine phena, the first five phena were found to belong to the genus Flavobacterium.

In a study by Owen and Holmes, the conclusion was drawn that Hayes' phenon 1 corresponded closely to CDC Group IIb; and some members of this taxon were later attributed to *C. indologenes* and *C. gleum*.

The 53 strains in this phenon were isolated from the following sources: raw beef carcasses, raw lamb carcasses, raw pig carcasses, raw eviscerated chicken carcasses, raw milk, water from rivers or streams and soil. Consequently Chryseobacterium strains clearly are found in a variety of meat products, but no mention was made about their spoilage role in these products.

Chryseobacterium gleum and *C. indologenes* are often present on raw meat. *Chryseobacterium vrystaatense* was present on raw chicken, obtained from a chicken-processing plant.

SPOILAGE CAUSED BY FLAVOBACTERIA

Flavobacteria have been associated with spoilage of food, but information about the incidence and role of flavobacteria in food deterioration is difficult to obtain, mainly due to the history of faulty classification or reclassification of these organisms. They are, however,

Food Spoilage

accepted as common contaminants of protein-rich foods and under refrigerated storage, they are in competition with the pseudomonads. Undesirable flavours and odours, possible slime production and/or toxic metabolic end products are detrimental and apart from an economical loss to industry and consumers, may also have a health impact on consumers. Even if the spoilage bacteria are not pathogenic *per sé*, changes in the biochemical status of stored food due to deterioration by such bacteria, may make conditions favourable for other bacteria, or even pathogens, to grow in.

Studies on the proteolytic activities of flavobacteria have indicated that flavobacteria may possibly produce pasteurisation resistant extracellular enzymes and that they may in this way contribute to the psychrotrophic spoilage of milk and dairy products.

Although psychrotrophs secrete other enzymes with spoilage potential, e.g., glycosidases, the most important enzymes from the viewpoint of food spoilage are extracellular proteinases, lipases, and phospholipases on which this review will concentrate.

Proteolytic activity

All enzymes that catalyse hydrolysis of proteins to peptones, polypeptides, and amino acids, are called proteolytic enzymes. These enzymes hydrolytically cleave the peptide linkage with the formation of a free amino and carboxylic acid group. Animal proteinases include such enzymes as pepsin, rennet, trypsin, chromotrypsin, and cathepsin. Continued proteolysis results in putrid off-flavours associated with lower molecular-weight degradation products such as ammonia, amines, and sulphides. Proteinase production by psychrotrophs is normally at a maximum in the late exponential or stationary phase of growth.

Bitter peptides are normally characterised by large numbers of hydrophobic amino acids. Proteases produced by psychrotrophs have been shown to hydrolyse casein, but whey proteins were more resistant against hydrolysis.

The optimum pH and temperature for protease production depends on the species and strain. The most common proteolytic activity in milk was reported as clotting. Roussis *et al.* found that Flavobacterium MTR3 proteinases were active at 32-45°C, and exhibited considerable activity at 7°C. The enzyme was active at pH 6.0-8.0, and exhibited considerable activity at pH 6.0 in the presence of 4 per cent NaCl.

Lipolytic activity

Lipolytic enzymes can be defined as carboxylesterases that hydrolyse acylglycerols. Most bacterial lipases are extracellular and are produced during the late log and early stationary phases of growth. True lipases act on insoluble substrates such as micelles in emulsion or surface monolayers. Lipolysis is known to contribute both desirable and undesirable flavours to dairy products, initially through hydrolysis of milk triacylglycerols.

Short-chain fatty acids, such as butyric acid, caproic acid and caprylic acid, impart sharp and tangy flavours. Medium-chain fatty acids, such as capric and lauric acid tend to impart a soapy taste, while longchain fatty acids, such as myristic acid, palmitic acid and stearic acid, contribute little to flavour.

Unsaturated fatty acids released during lipolysis are susceptible to oxidation and the concomitant formation of aldehydes and ketones, which give rise to off-flavours described as "oxidised card-boardy" or metallic. 2The lipases from many of the psychrotrophic bacteria are remarkably heat stable and may, therefore, contribute to lipolysis in dairy products, even when they are heat treated.

The microbial lipases can attack intact fat globules and may cause lipolysis without any prior activation. Other unpleasant flavours, such as "rancid, butyric, bitter, unclean, soapy and astringent" in milk and milk products, have also been attributed to lipolysis. In general, flavobacteria are less well-known for lipase production.

However, significant lipase production by some Flavobacterium strains have been reported by some researchers. Optimal temperature for extracellular Gram negative bacterial lipases is found in the temperature range of 30 to 40°C.

Bacterial lipases appear to be very stable at temperatures below 8°C. The optimum pH of most extracellular Gram negative bacterial lipases appears to be at neutral or alkaline pH values between seven and nine.

It has been suggested that the optimum pH depends upon the nature of the substrate, the buffer solution, and other external conditions.

Phospholipase C production

The production of phospholipases, especially type C or lecithinase by some Flavobacterium strains have been reported by some researchers.

Phospholipases are a complex group of enzymes which act on phospholipids.

Most bacterial phospholipases are of the C-type, that is, they hydrolyse phospholipids to diglycerides and substituted phosphoric acid.

There are at least two subclasses of phospholipase C:
- Those that hydrolyse phosphatidylcholine, phosphatidylethanolamine, or phosphatidylserine and
- Those that hydrolyse phosphatidylinositol.

Phospholipases are potentially important in milk and milk products because of their ability to degrade the phospholipids of the milk fat globule membrane, thereby increasing the susceptibility of the milk fat to lipolytic attack. Extracellular phospholipases produced by psychrotrophs growing in stored raw milk have the potential to exaggerate the problem of rancidity.

5

Scope of Food Microbiology

Microbiology is the science which includes the study of the occurrence and significance of bacteria, fungi, protozoa and algae which are the beginning and ending of intricate food chains upon which all life depends.

These food chains begin wherever photosynthetic organisms can trap light energy and use it to synthesize large molecules from carbon dioxide, water and mineral salts forming the proteins, fats and carbohydrates which all other living creatures use for food. Within and on the bodies of all living creatures, as well as in soil and water, micro-organisms build up and change molecules, extracting energy and growth substances.

They also help to control population levels of higher animals and plants by parasitism and pathogenicity. When plants and animals die, their protective antimicrobial systems cease to function so that, sooner or later, decay begins liberating the smaller molecules for re-use by plants. Without human intervention, growth, death, decay and regrowth would form an intricate web of plants, animals and micro-organisms, varying with changes in climate and often showing apparently chaotic fluctuations in populations of individual species, but inherently balanced in numbers between producing, consuming and recycling groups.

In the distant past, these cycles of growth and decay would have been little influenced by the small human population that could be supported by the hunting and gathering of food. From around 10000 BC, however, the deliberate cultivation of plants and herding of animals started in some areas of the world.

The increased productivity of the land and the improved nutrition that resulted led to population growth and a probable increase in the average life span. The availability of food surpluses also liberated some

Scope of Food Microbiology

from daily toil in the fields and stimulated the development of specialized crafts, urban centres, and trade—in short, civilization.

BASIC OF MICROBIOLOGY

Let's review, in general, the microbiology basics that you learned in Veterinary School. As a FSIS-PHV public health official it is important for you to understand the dynamics of those pathogens of concern to the food industry and consumers. As you know microbiology is defined as the science that deals with the study of microorganisms, including algae, bacteria, fungi, protozoa, and viruses. Specifically, bacteria are the most abundant of all organisms, they are unicellular, are relatively small ranging in size from 0.5- to 5.0 ìm, and for the most part they reproduce asexually. Although there are bacterial species capable of causing human illness and food spoilage, there are also beneficial species that are essential to good health and the environment.

Every bacterial species have specific nutritional requirements, temperature, humidity, etc., for energy generation and cellular biosynthesis. The bacterial cells divide at a constant rate depending upon the composition of the growth medium and the conditions of incubation and under favourable conditions, a growing bacterial population doubles at regular intervals ranging from about 15 minutes to 1 hour. This means that if we start with 1,000 cells with a generation time of 30 min. then after an hour we end with 4,000 cells.

The parametres affecting bacterial growth will be discussed. Bacteria are also known as prokaryotes because they don't possess nuclei; i.e., their chromosome is composed of a single closed double-stranded DNA circle. Structurally, a prokaryotic cell has three architectural regions: appendages in the form of flagella and pili; a cell envelope consisting of a capsule, cell wall and plasma or inner membrane; and a cytoplasmic region that contains the cell genome, ribosomes and various sorts of inclusions. Following is a brief discussion of these structural components.

- *Cell Envelope is Made of Three Layers*: Cytoplasmic membrane, the cell wall, and—in some bacterial species- an outer capsule. The role of the bacterial capsule is to keep the bacterium from drying, can serve as a virulence factor and as an antigen for identification, mediate adherence of cells to surface, and confer protection against engulfment and attack by antimicrobial agents of plants, animals, and the environment. Bacteria can be placed into two basic groups,

Gram-positive or Gram-negative, based on the profiles of the bacterial cell wall.
- Chromosome where the bacterium's genetic information is contained. It is a crucial tool for genetic fingerprinting.
- Cytoplasm is where the function for cell growth, metabolism, and replication are carried out. It is composed of water, enzymes, nutrients, metabolic wastes, and gases; it also contains the ribosomes, chromosomes, and plasmids. The cell envelope encases the cytoplasm and all its components.
- Flagella are hair-like structures that serve as propellers to help bacterium move towards nutrients and away from toxic chemicals. This structure can be found at either or both ends or all over the bacterium surface and serve as antigen for serotyping. Also, this organelle is a contributor for biofilm formation.
- Pili and fimbriae many species of bacteria have these small hair-like projections emerging from the outside cell surface. Its function is to assist in attaching to other cells and surfaces. Specialized pili are used for passing nuclear material between bacterial cells.
- Plasmid short length of extra-chromosomal genetic structure which are carried by many strains of bacteria. They are not involved in reproduction but replicate independently of the chromosome and are instrumental in the transmission of special properties, such as antibiotic drug resistance, resistance to heavy metals, and virulence factors necessary for infection of animal and human hosts. Plasmids are extremely useful tools in the area of genetic engineering.
- Ribosomes these are organelles that translate the genetic code DNA to amino acids which are the building blocks of proteins. They are also an important tool in the fields of molecular biology and genetics.
- Spores produced by some species and they are resistant to hostile conditions such as heat and drying. They serve as survival mechanisms when environmental conditions are not suitable for growth and replication.

The cell wall of bacteria is dynamic and extremely important for several reasons:
- They are an essential structure for viability; protects the cell protoplast from mechanical damage and from osmotic rupture or lysis.

Scope of Food Microbiology

- They are composed of unique components found nowhere else in nature.
- They are one of the most important sites for attack by antibiotics.
- They provide ligands for adherence and receptor sites for drugs or viruses.
- They cause symptoms of disease in humans and animals.
- They provide for immunological distinction and immunological variation among strains of bacteria.
- They can be modified to protect the cell against harsh environmental conditions like heat, pH, anti-microbials, etc.

Cell wall composition varies widely amongst bacteria and is an important factor in bacterial species analyses and differentiation. The main functions are to give the cell its shape and surround the cytoplasmic membrane, protecting it from the environment. The profiles of the cell walls of bacteria, as seen with the electron microscope, make it possible to distinguish two basic types of bacteria as follows:

- *Gram-positive Bacteria*: The cell wall adjoining the inner or cytoplasmic membrane is thick, consisting of several layers of peptidoglycan, also known as murein. Intertwine within the cell wall are polymers composed of glycerol, phosphates, and ribitol, know as teichoic acids. In general, Gram-positive bacteria produce extracellular substances that typically account for most of the virulence factors and this is showed by *Staphylococcus aureus*.
- *Gram-negative Bacteria*: The cell wall adjoining the inner membrane is relatively thin and is composed of a single layer of peptidoglycan surrounded by a membranous structure called the outer membrane. The outer membrane of Gram-negative bacteria invariably contains a unique component, lipopolysaccharide which is toxic to animals. This outer membrane is usually thought of as part of the cell wall. The pathogenesis and virulence properties of Gram-negative bacteria are far more complex including outer membrane components as well as the production of extracellular substances which can be showed by *E. coli* O157:H7.

It may be advantageous for epidemiological purposes to identify a particular bacterial strain by serotyping, which is a useful tool to accomplish this goal. There are components in the cell envelope that serves as antigens for serotyping, therefore, serotyping is based on the ability of the bacteria to agglutinate antibodies specific for those antigens.

Following is a brief description regarding to the serotyping of those pathogens of public health concern. Serotyping of Gram-negative bacteria consist of the immunoreactivity of three classes of antigens: the O-antigen, H-antigen, and the K-antigen surface profiles. The O-antigen is a polysaccharide which is a polymer of O-subunits, composed of 4-6 sugar residues, attached to the lipid A-core polysaccharide portion of the LPS molecule.

Differences in the immunoreactivity of antibodies with the O-antigen result from the variation in the sugar components and/or covalent linkages between the O-subunit. On the other hand, the H-antigen is the filamentous portion of the flagella which is composed of protein subunits called flagellin. The antigenically variable portion of flagellin determines the H serotype as determined by H antiserum. Finally, the K-antigens are the somatic or surface antigens that occur as envelopes, sheaths, or capsules. They act as masking antigens for the O-antigen, inhibiting agglutination of living cell suspensions in O antiserum. A specific combination of O- and H-antigens defines what is known as the serotype and/or serogroups of a bacterial isolate.

The serotype and serogroups in particular species provide identifiable chromosomal markers that correlate with specific bacterial virulent clones. More than 2,500 Salmonella serotypes have been described and reported; examples are *S. Enteritidis* and *S. Newport* which belong to Serogroup D and B, respectively. In *E. coli*, a total of 170 different O-antigens and 55 H-antigens, defining the isolate serotype, have been identified; a well known example is *E. coli* O157:H7 serotype which is part of the enterohemorrhagic serogroup. Likewise, the serotyping of Gram-positive bacteria is based on the combination of somatic and flagellar antigens. Although serological confirmation is not necessary for regulatory identification of *L. monocytogenes*, it is useful for determining the prevalence of specificserotypes in epidemiological studies and for environmental recontamination tracking. Strains of *L. monocytogenes* can be assigned to 13 different serotypes, based on their combination of O- and H-antigens. While all of them are considered to be potentially pathogenic, most human clinical isolates belong to three serotypes 1/2a, 1/2b, and 4b. It is evident that bacteria are a complex system with the capability to adapt and survive to adverse environmental conditions. This explains, in part, why there are some microorganisms that are very difficult to eliminate, why other becomes pathogenic, and why other develops resistance towards antibiotics or antimicrobial interventions. In

Scope of Food Microbiology

slaughter as well as in the processing establishments there are bacterial species associated with particular meat and poultry products, including the environment.

THE HISTORY OF MICROBIOLOGY

A look at the history of microbiology will help you to understand the contributions of those who have come before. This perspective will hopefully give you an appreciation of their efforts and put the body of knowledge we will examine in the context of history. Keep in mind that microbiology is a relatively young science. It was only 130 years ago that it became possible to seriously study microorganisms in the laboratory, with most of our understanding of microbes coming in the last 60 years. The history of microbiology, like all human history, is not a catalogue of linear progress, but is more of an interweaving of the careers of bright individuals and their insights.

Each new discovery relied on previous ones and in turn spawned further enquiry. A web of interdependent concepts evolved over time through the work of scientists in many related disciplines and nations. Often the research of one individual impacted the efforts of another studying a completely different problem. Keep this in mind as you look at this history.

Below we present several journeys through this web, mentioning some individuals who were particularly important in the progression. This history reflects our view of important events of the past, but is by no means comprehensive.

We will first look at the development of the techniques for handling microorganisms, since everything else in microbiology depends upon these procedures. Next, we will examine how these techniques helped to settle an old debate, the question of spontaneous generation. Then, we will look at the history of infectious disease. The science of microbiology had its most significant early impact on human health, uncovering the cause of the major killers of the day, and then methods to treat them. As microbiology matured, scientists began to look at what non-pathogenic microbes were doing in the environment and we will look a bit at the history of general microbiology. Finally, the chapter will end with an examination of the events that lead to the understanding of life at the molecular level and the profound impact this has had on microbiology and on society in general.

Importance of Microscopes

It took the microscope to expose their tiny world and that instrument has been linked to microbiology ever since. In 1664, Robert Hooke devised a compound microscope and used it to observe fleas, sponges, bird feathers, plants and molds, among other items. His work was published in *Micrographia* and became a popular and widely read book at the time. Several years later Anton van Leeuwenhoek, a fabric merchant and amateur scientist (or "natural philosopher" as such people called themselves), became very adept at grinding glass lenses to make telescopes and microscopes. While crude by modern standards, his were a technical marvel for the time, able to magnify samples greater than 200-fold. They also produced clearer images than the compound microscopes of the time. By peering through his microscope, Leeuwenhoek observed tiny organisms or "wee animacules" as he called them.. He spent months looking at every kind of sample he could find and eventually submitted his observations in a letter to the Royal Society of London, causing a sensation. Hooke was asked to confirm the findings of Leeuwenhoek and his affirmative assessment garnered them wide acceptance. Surprisingly the work of these two scientists was not followed up for almost 200 years.

Human societies had neither the technical prowess nor the inclination to develop the science of microorganisms. It was not until the rise of the industrial revolution that governments and people dedicated the financial and physical resources to understand these small inhabitants of our world. With the development of better microscopes in the 19th Century, scientists returned to an examination of microorganisms. After finishing his education, Ferdinand Julius Cohn was able to convince his father to lay down the large sum necessary to purchase a microscope for him, one better than that available at the University in Breslau, then part of Germany.

He used it to carefully examine the world of the microbe and made many observations of eukaryotic microorganisms and bacteria. His landmark papers on the cycling of elements in nature was published in *Ueber Bakterien* in 1872 and a microbial classification scheme including descriptions of *Bacillus* were published in the first volume of a journal he founded, *Beitraege zur Biologie der Planzen*. Cohn's work with microscopes popularized their use in microbiology. This and his other work inspired many other scientists to examine microbes. Cohn's encouragement of Robert Koch, a German physician by training, began the field of medical microbiology.

Scope of Food Microbiology

HUMAN EFFECTED BY MICROBES

"Ancient" diseases continue to be a problem where nutrition and sanitation are poor, and emerging diseases such as Acquired Immunodeficiency Syndrome (AIDS) are even more dangerous for such populations. The Centres for Disease Control and Prevention (the U.S. government agency charged with protecting the health and safety of people) estimates that about 9 per cent of adults between the ages of 18-49 in Sub-Saharan Africa are infected with HIV. AIDS is only one of a number of new diseases that have emerged.

Many of these diseases have no known cure. Influenza and pneumonia are leading killers of the elderly even in the U. S. and other developed nations.

Even the common cold causes illness and misery for almost everyone and drains the productivity of all nations. Many of the new diseases are viral in nature, making them notoriously difficult to treat. Disease due to food-borne pathogens also remains a problem, largely because of consumption of improperly processed or stored foods. Understanding the sources of contamination and developing ways to limit the growth of pathogens in food is the job of food microbiologists. New infections continually appear.

Having an available food source to grow on (humans) inevitably results in a microorganism that will take advantage. Some of these feeders will interfere with our own well being, causing disease. Surprisingly, many diseases that were previously thought to have only behavioural or genetic components have been found to involve microorganisms. The clearest case is that of ulcers, which was long thought to be caused by stress and poor diet. However the causative agent is actually a bacterium, *Helicobacter pylori*, and many ulcers can be cured with appropriate antibiotics.

Work on other non-infectious diseases such as heart disease, stroke and some autoimmune diseases also suggest a microbial component that triggers the illness. Finally, some pathogenic microbes that had been "controlled" through the use of antibiotics are beginning to develop drug resistance and therefore re-emerge as serious threats in the industrialized world as well as developing nations. Tuberculosis is an illness that was on the decline until the middle 80's. It has recently become more of a problem, partly due to drug resistance and partly due to a higher population of immunosuppressed individuals from the

AIDS epidemic. *Staphylococcus aureus* strains are emerging that are resistant to many of the antibiotics that were previously effective against them. These staph infections are of great concern in hospital settings around the world. Understanding both familiar killers and new pathogens will require an understanding of their biology, and thus an understanding of the field of microbiology.

BENEFITS OF MICROBES

Significant resources have been spent to understand and fight disease-causing microorganisms from the beginning of microbiology. You may be surprised to learn that only a small faction of microbes are involved in disease; many other microbes actually enhance our well-being.

The harmless microbes that live in our intestines and on our skin actually help us fight off disease. They actively antagonize other bacteria and take up space, preventing potential pathogens from gaining a foothold on our bodies.

Indeed, like all large organisms, humans have entire communities of microorganisms in their digestive systems that contribute to their overall health. The microbial community in humans not only protects us from disease, but also provides needed vitamins, such as B_{12}. Human health and nutrition also depends on healthy farm animals. Cows, sheep, horses and other ruminant animals utilize their microbial associates to degrade plant material into useful nutrients.

Without these bacteria inside ruminants, growth on plant material would be impossible. In contrast to humans ruminant animals have complex stomachs that harbour large numbers of microorganisms. These microbes degrade the plant stuff eaten by the animal into usable nutrients. Without the assistance of the microbes, ruminant animals would not be able to digest the food they eat. Commercial crops are also central to human prosperity, and much of agriculture depends upon the activities of microbes. For example, an entire group of plants, the legumes, forms a cooperative relationship with certain bacteria.

These bacteria convert nitrogen gas to ammonia for the plant, an important nutrient that is often limiting in the environment. Microbes also serve as small factories, producing valuable products such as cheese, yogurt, beer, wine, organic acids and many other items. In conclusion, while it is less apparent to us, the positive role of microbes in human health is at least as important as the negative impact of pathogens. A

Scope of Food Microbiology

picture of nodulated leguminous plants. In this case pea plants. Nodules are visible on the roots of the plants in the left of the picture. The plants on the right were not inoculated with nodulating bacteria.

ENVIRONMENT EFFECTED BY MICROBES

The vast majority of life on this planet is microscopic. These teaming multitudes profoundly influence the make-up and character of the environment in which we live. Presently, we know very little about the microbes that live in the world around us because less than 2 per cent of them can be grown in the laboratory. Understanding which microbes are in each ecological niche and what they are doing there is critical for our understanding of the world. Microbes are the major actors in the synthesis and degradation of all sorts of important molecules in environments.

Cyanobacteria and algae in the oceans are responsible for the majority of photosynthesis on Earth. They are the ultimate source of food for most ocean creatures (including whales) and replenish the world's oxygen supply. Cyanobacteria also use carbon dioxide to synthesize all of their biological molecules and thus remove it from the atmosphere.

Since carbon dioxide is a major greenhouse gas, its removal by cyanobacteria affects the global carbon dioxide balance and may be an important mitigating factor in global warming. In all habitats, microorganisms make nutrients available for the future growth of other living things by degrading dead organisms. Microbes are also essential in treating the large volume of sewage and wastewater produced by metropolitan areas, recycling it into clean water that can be safely discharged into the environment.

Less helpfully (from the view of most humans), termites contain microorganisms in their guts that assist in the digestion of wood, allowing the termites to extract nutrients from what would otherwise be indigestible. Understanding of these systems helps us to manage them responsibly and as we learn more we will become ever more effective stewards. Energy is essential for our industrial society and microbes are important players in its production. A significant portion of natural gas comes from the past action of methanogens (methane-producing bacteria). Numerous bacteria are also capable of rapidly degrading oil in the presence of air and special precautions have to be taken during the drilling, transport and storage of oil to minimize their impact. In the future, microbes may find utility in the direct production of energy. For example, many landfills and sewage treatment plants capture the methane produced by methanogens to power turbines that produce electricity.

Excess grain, crop waste and animal waste can be used as nutrients for microbes that ferment this biomass into methanol or ethanol. These biofuels are presently added to gasoline and thus decreasing pollution. They may one-day power fuel cells in our cars, causing little pollution and having water as their only emission. Finally, We are increasingly taking advantage of the versatile appetite of bacteria to clean up environments that we have contaminated with crude oil, polycholrinated biphenyls (PCBs) and many other industrial wastes. This process is termed bioremediation and is a cheap and increasingly effective way of cleaning up pollution.

MICROBES AROUND US

Microorganisms will grow on simple, cheap medium and will often rise to large populations in a matter of 24 hours. It is easy to isolate their genomic material, manipulate it in the test tube and then place it back into the microbe. Due to their large populations it is possible to identify rare events and then, with the use of powerful selective techniques, isolate interesting bacterial cells and study them. These advantages have made it possible to test hypothesis rapidly. Using microbes scientists have expanded our knowledge about life. Below are a few examples. Microorganisms have been indispensable instruments for unlocking the secrets of life. The molecular basis of heredity and how this is expressed as proteins was described through work on microorganisms. Due to the similarity of life at the molecular level, this understanding has helped us to learn about all organisms, including ourselves. Some prokaryotes are capable of growing under unimaginably harsh conditions and define the extreme limits of where life can exist. Some species have been found growing at near 100 °C in hot springs and well above that temperature near deep-sea ocean vents.

Others make their living at near 0 °C in freshwater lakes that are buried under the ice of Antarctica. The ability of microbes to live under such extreme conditions is forcing scientists to rethink the requirements necessary to support life. Many now believe it is entirely possible that Jupiter's moon Europa may harbour living communities in waters deep below its icy crust. What may the rest of the universe hold? Until recently, while we could study specific types of bacteria, we lacked a cohesive classification system, so that we could not readily predict the properties of one species based on the known properties of others.

Scope of Food Microbiology

Visual appearance, which is the basis for classification of large organisms, simply does not work with many microbes because there are few distinguishing characteristics for comparison between species even under the microscope. However, analysis of their genetic material in the past 20 years has allowed such classification and spawned a revolution in our thinking about the evolution of bacteria and all other species. The emergence of a new system organizing life on Earth into three domains is attributable to this pioneering work with microorganisms. The fruits of basic research on microbes has been used by scientists to understand microbial activity and therefore to shape our modern world.

Human proteins, especially hormones like insulin and human growth factor, are now produced in bacteria using genetic engineering. Our understanding of the immune system was developed using microbes as tools. Microorganisms also play a role in treating disease and keeping people healthy. Many of the drugs available to treat infectious disease originate from bacteria and fungi. One last recent role of microbes in informing us about our world has been the tools they provide for molecular biology. Enzymes purified from bacterial strains are useful as tools to perform many types of analyses. Such analyses allow us to determine the complete genome sequence of almost any organism and manipulate that DNA in useful ways.

We now know the entire sequence of the human genome, with the exception of regions of repetitive DNA, and this will hopefully lead to medical practices and treatments that improve health. We also know the entire genome sequence of many important pathogens. Analysis of this data will eventually lead to an understanding of the function of critical enzymes in these microbes and the development of tailor-made drugs to stop them. The tools of molecular biology will also affect agriculture.

For example, we now know the complete genome sequence of the plant *Arabidopsis* (a close relative of broccoli and cauliflower). This opens a new avenue to a better understanding of all plants and hopefully improvements in important crops. Microbes have a profound impact on every facet of human life and everything around us. Pathogens harm us, yet other microbes protect us. Some microbes are pivotal in the growth of food crops, but others can kill the plants or spoil the produce.

Bacteria and fungi eliminate the wastes produced in the environment, but also degrade things we would rather preserve. Clearly they effect many things we find important as humans. In the remainder of this chapter we take a look at how scientists came to be interested in microbes and follow a few important developments in the history of microbiology.

SPECTRUM OF MICROBIOLOGY

Like all other living things, microorganisms are placed into a system of classification. Classification highlights characteristics that are common among certain groups while providing order to the variety of living things. The science of classification is known as taxonomy, and taxon is an alternative expression for a classification category. Taxonomy displays the unity and diversity among living things, including microorganisms. Among the first taxonomists was Carolus Linnaeus. In the 1750s and 1760s, Linnaeus classified all known plants and animals of that period and set down the rules for nomenclature.

Classification schemes

The fundamental rank of the classification as set down by Linnaeus is the species. For organisms such as animals and plants, a species is defined as a population of individuals that breed among themselves. For microorganisms, a species is defined as a group of organisms that are 70 per cent similar from a biochemical standpoint.

In the classification scheme, various species are grouped together to form a genus. Among the bacteria, for example, the species *Shigella boydii* and *Shigella flexneri* are in the genus *Shigella* because the organisms are at least 70 per cent similar. Various genera are then grouped as a family because of similarities, and various families are placed together in an order. Continuing the classification scheme, a number of orders are grouped as a class, and several classes are categorized in a single phylum or division. The various phyla or divisions are placed in the broadest classification entry, the kingdom. Numerous criteria are used in establishing a species and in placing species together in broader classification categories. Morphology and structure are considered, as well as cellular features, biochemical properties, and genetic characteristics. In addition, the antibodies that an organism elicits in the human body are a defining property. The nutritional format is considered, as are staining characteristics.

Prokaryotes and eukaryotes

Because of their characteristics, microorganisms join all other living organisms in two major groups of organisms: prokaryotes and eukaryotes.

Scope of Food Microbiology

Bacteria are prokaryotes because of their cellular properties, while other microorganisms such as fungi, protozoa, and unicellular algae are eukaryotes. Viruses are neither prokaryotes nor eukaryotes because of their simplicity and unique characteristics.

INDUSTRIAL MICROBIOLOGY/BIOTECHNOLOGY?

Industrial microbiology or microbial biotechnology is the application of scientific and engineering principles to the processing of materials by microorganisms (such as bacteria, fungi, algae, protozoa and viruses) or plant and animal cells to create useful products or processes. The microorganisms utilized may be natural isolates, laboratory selected mutants or microbes that have been genetically engineered using recombinant DNA methods. The terms "industrial microbiology" and "biotechnology" are often one and the same. Areas of industrial microbiology include quality assurance for the food, pharmaceutical, and chemical industries. Industrial microbiologists may also be responsible for air and plant contamination, health of animals used in testing products, and discovery of new organisms and pathways. For instance, most antibiotics come from microbial fermentations involving a group of organisms called actinomycetes. Other organisms such as yeasts are used in baking, in the production of alcohol for beverages, and in fuel production (gasohol). Additional groups of microorganisms form products that range from organic acids to enzymes used to create various sugars, amino acids, and detergents. For example, the sweetener aspartame is derived from amino acids produced by microorganisms.

Industrial Application of Microbes

Microbes have been used to produce products for thousands of years. Even in ancient times, vinegar was made by filtering alcohol through wood shavings, allowing microbes growing on the surfaces of the wood pieces to convert alcohol to vinegar. Likewise, the production of wine and beer uses another microbe — yeast — to convert sugars to alcohol. Even though people did not know for a long time that microbes were behind these transformations, it did not stop them from making and selling these products.

Both of these are early examples of biotechnology — the use of microbes for economic or industrial purposes. This field advanced considerably with the many developments in microbiology, such as the invention of microscope. Once scientists learned about the genetics of microbes, and how their cells produce proteins, microbes could also be altered to function in many new, and useful, ways. This sparked the application of biotechnology to many industries, such as agriculture, energy and medicine.

Genetic Engineering of Microbes

Genetic information in organisms is stored in their DNA. This molecule holds instructions for how the organism looks and functions. DNA is broken into sections called genes, each of which contains the template for a single protein molecule. Proteins serve as building blocks for the cell, and also carry out other activities. By studying microbes, scientists learned how to cut pieces out of a DNA molecule, and move them to another part. This changes how the cell looks or acts. Scientists can also take genes from one organism and insert them into the DNA of another. This gives the organism entirely new abilities.

This type of genetic engineering — the altering of an organism's genetic information — has enabled scientists to use microbes as tiny living factories. One example of this is the production of insulin. In humans, the pancreas creates a protein called insulin that regulates glucose — sugar — levels in the blood. People with one type of diabetes cannot produce insulin, so they inject it into their blood throughout the day. To produce cheaper insulin, scientists inserted the human gene for insulin into the DNA of a common intestinal bacterium. This change enabled the bacterium to produce a new product — human insulin.

Food and Agriculture and Microbes

As with the production of vinegar, microbes are used widely in the agricultural and food industries. Bacteria are used in the production of many food products, such as yogurt, many types of cheese and sauerkraut. Farmers also use a bacterium that produces a natural fertilizer. This type of bacterium is normally associated with bean plants, growing in nodules on the roots in a symbiotic — mutually beneficial — relationship. The bacterium converts nitrogen gas in the air to a form that plants can use — like fertilizer. By adding bacteria to the soil, farmers can increase the productivity of the plants.

Genetic engineering can also be used to produce plants with new abilities, such as enhanced resistance to pesticides, or increased nutritional content. In this case, microbes are used to insert new genes into the DNA of the plants. This results in genetically modified — GM — foods. Humans

Scope of Food Microbiology

have long modified the genetics of agricultural plants and animals by breeding them to enhance specific traits. Genetical engineering, however, allows scientists to add genes that exist in totally unrelated organisms.

Energy and Microbes

During vinegar production with wood chips, bacteria grow on the surface of the wood, forming what is called a biofilm. Bacteria attached to a surface like this can produce many compounds, as well as block the flow of a fluid. The latter behaviour has been used to increase the amount of oil extracted from an oil field. Bacteria growing in the wells block areas that are more open. When water is then pumped into the ground, the biofilms drive the water into other areas that still contain oil. This then forces the oil to the surface.

Microbes can also be used to create fuels directly. Certain bacteria ferment glycerol to form ethanol, a biofuel that can be used in automobiles. The glycerol is a byproduct of biodiesel production, but it is more valuable if converted to fuel. With genetic engineering, microbes can also be altered to produce fuels that they don't usually make. One company has modified the DNA of yeast to create biofuel from sugarcane feedstock. The challenge to all of these methods is creating a process that produces fuels more easily — and cheaply — than conventional methods.

Crime and Security and Microbes

Certain types of bacteria thrive in high temperatures. These extremophiles — organisms that prefer extreme conditions — have cell components designed to withstand heat. One of these is a bacterium, Thermus aquaticus, that lives in hot springs and near thermal vents. It contains an enzyme that is involved in the copying of DNA inside the cell. This type of enzyme occurs in other organisms, but the one from T. aquaticus can withstand higher temperatures. Scientists use this enzyme to multiply very small amounts of DNA, such as from samples found at crime scenes. Other techniques are used to identify disease-causing microbes released by terrorists. The microbes can be identified from their DNA. These tests are extremely sensitive, and can find the DNA equivalent of a drop of water in a swimming pool. The U.S. Postal used microbe-detection techniques after letters contaminated with a dangerous microbe — anthrax — were sent through the mail. The tests identified the microbe as coming from the same source, meaning that a single person sent all of the letters.

Medical Application of Microbes

In addition to vaccines and antibiotics, microbes have been essential for many important contributions to medicine. Like diabetes, many diseases can be treated with compounds derived from microbes: cystic fibrosis, cancer, growth hormone deficiency and hepatitis B. In addition, genetic methods that were first developed in microbes now allow scientists to study genetic diseases in humans. This has resulted in the ability to test fetuses for genetic diseases.

There have also been research studies of gene therapy in humans. This technique uses a microbe — often a virus — to insert new genes into cells. In theory, this could correct a condition caused by a genetic disease. Microbial genetics has also led to the ability to determine the sequence of DNA more rapidly, like reading a book. With this information, scientists can look for genes in individuals that cause — or contribute to — diseases.

MICRO-ORGANISMS AND FOOD

The foods that we eat are rarely if ever sterile, they carry microbial associations whose composition depends upon which organisms gain access and how they grow, survive and interact in the food over time. The micro-organisms present will originate from the natural microflora of the raw material and those organisms introduced in the course of harvesting, slaughter, processing, storage and distribution.

The numerical balance between the various types will be determined by the properties of the food, its storage environment, properties of the organisms themselves and the effects of processing. In most cases this microflora has no discernible effect and the food is consumed without objection and with no adverse consequences. In some instances though, micro-organisms manifest their presence in one of several ways:

- They can cause spoilage.
- They can cause foodborne illness.
- They can transform a food's properties in a beneficial way—food fermentation.

Food Preservation

From the earliest times, storage of stable nuts and grains for winter provision is likely to have been a feature shared with many other animals but, with the advent of agriculture, the safe storage of surplus production assumed greater importance if seasonal growth patterns were to be used most effectively. Food preservation techniques based on sound, if then unknown, microbiological principles were developed empirically to arrest or retard the natural processes of decay. The staple foods for most parts of the world were the seeds-rice, wheat, sorghum, millet, maize, oats and barley—which would keep for one or two seasons if adequately dried, and it seems probable that most early methods of food preservation depended largely on water activity reduction in the form of solar drying, salting, storing in concentrated sugar solutions or smoking over a fire.

The Industrial Revolution which started in Britain in the late 18th century provided a new impetus to the development of food preservation techniques. It produced a massive growth of population in the new industrial centres which had somehow to be fed; a problem which many thought would never be solved satisfactorily.

Such views were often based upon the work of the English cleric Thomas Malthus who in his *Essay on Population* observed that the inevitable consequence of the exponential growth in population and the arithmetic growth in agricultural productivity would be over-population and mass starvation. This in fact proved not to be the case as the 19th century saw the development of substantial food preservation industries based around the use of chilling, canning and freezing and the first large scale importation of foods from distant producers. To this day, we are not free from concerns about over-population. Globally there is sufficient food to feed the world's current population. World grain production has more than managed to keep pace with the increasing population in recent years and the World Health Organization's Food and Agriculture Panel consider that current and emerging capabilities for the production and preservation of food should ensure an adequate supply of safe and nutritious food to the world's population.

There is however little room for complacency. Despite overall sufficiency, it is recognized that a large proportion of the population is malnourished. The principal cause of this is not insufficiency, however, but poverty which leaves an estimated one-fifth of the world's population without the means to meet their daily needs. Any long-term solution to this must lie in improving the economic status of those in the poorest countries and this, in its train, is likely to bring a decrease in population growth rate similar to that seen in recent years in more affluent countries.

In any event, the world's food supply will need to increase to keep pace with population growth and this has its own environmental and social costs in terms of the more intensive exploitation of land and sea resources.

One way of mitigating this is to reduce the substantial pre and post-harvest losses which occur, particularly in developing countries where the problems of food supply are often most acute.

It has been estimated that the average losses in cereals and legumes exceed 10% whereas with more perishable products such as starchy staples and vegetables the figure is more than 20%—increasing to an estimated 25% for highly perishable products such as fish. In absolute terms, the US National Academy of Sciences has estimated the losses in cereals and legumes in developing countries as 100 million tonnes, enough to feed 300 million people.

Clearly reduction in such losses can make an important contribution to feeding the world's population. While it is unrealistic to claim that food microbiology offers all the answers, the expertise of the food microbiologist can make an important contribution. In part, this will lie in helping to extend the application of current knowledge and techniques but there is also a recognized need for simple, low cost, effective methods for improving food storage and preservation in developing countries.

Problems for the food microbiologist will not however disappear as a result of successful development programmes. Increasing wealth will lead to changes in patterns of food consumption and changing demands on the food industry. Income increases among the poor have been shown to lead to increased demand for the basic food staples while in the better-off it leads to increased demand for more perishable animal products.

To supply an increasingly affluent and expanding urban population will require massive extension of a safe distribution network and will place great demands on the food microbiologist.

Food Safety

In addition to its undoubted value, food has a long association with the transmission of disease. Regulations governing food hygiene can be found in numerous early sources such as the Old Testament, and the writings of Confucius, Hinduism and Islam. Such early writers had at best only a vague conception of the true causes of foodborne illness and many of their prescriptions probably had only a slight effect on its incidence. Even today, despite our increased knowledge, foodborne disease is perhaps the most widespread health problem in the contemporary world and an important cause of reduced economic productivity.

The available evidence clearly indicates that biological contaminants are the major cause. The various ways in which foods can transmit illness,

Scope of Food Microbiology

the extent of the problem and the principal causative agents. Microbes can however play a positive role in food. They can be consumed as foods in themselves as in the edible fungi, mycoprotein and algae. They can also effect desirable transformations in a food, changing its properties in a way that is beneficial. Food microbiology is unashamedly an applied science and the food microbiologist's principal function is to help assure a supply of wholesome and safe food to the consumer. To do this requires the synthesis and systematic application of our knowledge of the microbial ecology of foods and the effects of processing to the practical problem of producing, economically and consistently, foods which have good keeping qualities and are safe to eat.

GROWTH OF MICROORGANISMS

There are basically two parametres that affect the growth of microorganisms in food products, extrinsic and intrinsic. Extrinsic parametres are those properties of the environment that exist outside of the food product which affect both the foods and their microorganisms. On the other hand, intrinsic parametres, are properties that exist as part of the food product itself, for example, tissues are an inherent part of the animal that may, under a set of conditions, promote microbiological growth.

Following is a list of these parametres that either may result in multiplication or inhibition of microbial growth in meat, poultry, or egg product. Examples of intrinsic parametres are:

- *pH*: It has been well established that most microorganisms grow best at pH values around 7.0, whereas few grow below a pH of 4.0. Bacteria tend to be more fastidious in their relationships to pH than molds and yeasts, with the pathogenic bacteria being the most fastidious. Most of the meats have a final pH of about 5.6 and above; this makes these products susceptible to bacteria as well as to mold and yeast spoilage.
- *Moisture Content*: One of the oldest methods of preserving foods is drying or desiccation. The preservation of foods by drying is a direct consequence of removal or binding of moisture, without which microorganisms do not grow. It is now generally accepted that the water requirements of microorganisms should be described in terms of water activity in the environment. Basically, the water molecules are loosely oriented in pure liquid water and can easily rearrange. When a solute is added to water, the water molecules orient themselves on the surface of the solute, in

this case the Na^+ and Cl^- ions, and the properties of the solution change dramatically. Therefore, the microbial cell must compete with solute molecules for free water molecules. The water activity of pure water is 1.00; the addition of solute decreases a_w to less than 1.00. Most foodborne pathogenic bacteria require a_w greater than 0.9, however, *Staphylococcus aureus* may grow in a_w as low as 0.86.

- *Oxidation-reduction Potential*: Microorganisms display varying degrees of sensitivity to the oxidation-reduction potential of their growth medium or environment. Aerobic microorganisms require more oxidized environments versus anaerobic organisms which require more reduced environments.
- *Nutrient Content*: In order to grow and function normally, the microorganisms of concern in the food industry require the following: water, source of energy, source of nitrogen, vitamins and related growth factors, and minerals.
- *Antimicrobial Constituents*: The stability of some foods against attack by microorganisms is due to the presence of certain naturally occurring substances that have been shown to have antimicrobial activity. Nisin and other bacteriocins are good examples.
- *Biological Structures*: The natural covering of some food sources provides excellent protection against the entry and subsequent damage by spoilage organisms. Examples of such protective structure are the hide, skin and feathers of animals.

Examples of extrinsic parametres are:

- *Storage Temperature*: Microorganisms, individually and as group, grow over a wide range of temperatures. It is important to know the temperature growth ranges for organisms of importance in foods as an aid in selecting the proper temperature for product storage. A helpful reference is the FDA's Food Code; it contains some recommendations for storage temperatures of product that are widely accepted in the scientific community.
- *Relative Humidity*: The relative humidity of the storage environment is important both from the standpoint of water activity within foods and the growth of microorganisms at the surfaces. Humidity can also be an important factor to consider when producing some types of product.
- *Presence/Concentration of Gases*: Carbon dioxide is the single most important atmospheric gas that is used to control

microorganisms in foods. It has been shown to be effective against a variety of microorganisms. Because of its effectiveness, CO_2 is used as one of the methods for modified-atmosphere packaging.

- *Presence/Activities of other Microorganisms*: The inhibitory effect of some members of the food microbiota on other microorganisms is well established. Some foodborne organisms produce substances that are either inhibitory or lethal to others. These include antibiotics, bacteriocins, hydrogen peroxide, and organic acids. General microbial interference is a phenomenon that refers to general nonspecific inhibition or destruction of one microorganism by other members of the same habitat or environment; the mechanism for this interference is not very clear. Some of the possibilities are: competition for nutrients; competition for attachment/adhesion sites; unfavourable alteration of the environment and/or combinations of these.

Isolation And Identification Of Pathogens

You know that FSIS is responsible for aseptically collecting samples to determine the presence of pathogens and Salmonella species just as to the regulations. Once these samples are received by the Agency's laboratory, how are they processed? When samples are received by any of the three field service laboratories it is first subject to a selective enrichment procedure, to favour growth of the desired organism, followed by an initial screening test for presumptive positives. The BAX® system is used as one of the initial screening test for the detection of Salmonella, *Listeria monocytogenes* and *Escherichia coli* O157:H7, and is based on the polymerase-chain reaction technology which has proven to be rapid and highly sensitive.

Thereafter, those found to be a screening positive will be further confirmed using immunological, biochemical, and molecular methods. Let's look at an example pertaining to the isolation and characterization of E. coli O157:H7 and/or O157:H7/NM from raw and ready-to-eat beef products.

The first step is to enrich the samples using an enrichment broth suitable for this pathogen followed by the screening test using the BAX® system in conjunction with alternative screening test. Those samples found to be positive are further processed by performing an immuno-magnetic separation using magnetic beads coated with O157-antibodies and plating an aliquot on a selective media; plates are then incubated for 18-24 hours at 35°C.

One or more typical colonies are tested with O157 antiserum and colonies that show agglutination are processed for confirmation by performing serological, biochemical, Shiga toxin assays, and genetic analyses. The time frame for reporting potential positive or screen negative result is two days; presumptive positive is 3 days; and confirm positives is 5-7 days.

The time frame for reporting the test results of these microorganisms is as follows:

- *L. monocytogenes*: for screen negative is 3 days; presumptive positive is 4-5 days; and confirmed positive is 5-8 days.
- *Salmonella spp*: for screen negative is 2 days; presumptive positive is 5 days; and confirmed positive is 7 days.

These results are then posted on LEARN to be accessible for the FSIS inspection personnel. Remember that, in the case of *E. coli* O157:H7 and *L. monocytogenes* in ready-to-eat products, presumptive positives reports are also posted so immediate action can be taken by the establishment concerning to the adulterated product.

6
Prevention of Spoilage in Milk

In the early days of development of the commercial dairy industry, milk was produced under much less sanitary conditions than are used today, and cooling was slow and inadequate to restrict bacterial growth. Developments during the first half of the twentieth century created significant reductions in the rate of spoilage of raw milk and cream, by making it possible for every-other-day pickup of milk from farms and shipments of raw milk over long distances with minimal increases in bacterial cell numbers.

Rapid cooling and quick use of raw milk are accepted as best practices and can affect the spoilage ability of *Pseudomonas* spp. present in milk. Pseudomonads that had been incubated in raw milk for 3 days at 7°C had greater growth rates and greater proteolytic and lipolytic activity than those isolated directly from the milk shortly after milking. As the quality of raw milk improved, so did that of pasteurized milk. Heating of milk to 62.8°C (145°F) for 30 min or to 71.7°C (161°F) for 15 skills the pathogenic bacteria likely to be of significance in milk as well as most of the spoilage bacteria.

However, processors learned that long shelf life of pasteurized fluid milk products requires a higher temperature treatment as well as prevention of contamination between the pasteurizer and the sealed package. In particular, it is imperative that filling equipment be sanitary and that the air in contact with the filler, the milk, and the containers be practically sterile. Whereas in the early to mid-twentieth century, milk was delivered daily to homes because of its short shelf life, today's fluid milk products are generally expected to remain acceptable for 14–21 days.

Pasteurization standards for several countries are listed in Table. A shelf life of 21 days and beyond can be attained with fluid milk products that have been heated sufficiently to kill virtually all of the vegetative bacterial cells and protected from recontamination. Ultra-pasteurized milk products, heated at or above 138°C for at least 2s, that have been packaged aseptically can have several weeks of shelf life when stored refrigerated. Ultra-high-temperature (UHT) treatment destroys most spores in milk, but *B. stearothermophilus* can survive.

Aseptic processing, as defined in the Grade A Pasteurized Milk Ordinance, means that the product has been subjected to sufficient heat processing to render it commercially sterile and that it has been packaged in a hermetically sealed container. These dairy foods are stable at room temperature. The addition of carbon dioxide to milk and milk products reduces the rates of growth of many bacteria. King and Mabbitt demonstrated improved keeping quality of raw milk by the addition of CO2. Loss and Hotchkiss found lowered survivor rates of both *P. fluorescens* and the spores of *B. cereus* during heating of milk containing up to 36 mM CO2.

Prevention of Spoilage in Cultured Dairy Products

Cultured products such as buttermilk and sour cream depend on a combination of lactic acid producers, the lactococci, and the leuconostocs to produce the desired flavour profile. Imbalance of the culture, improper temperature or ripening time, infection of the culture with bacteriophage, presence of inhibitors, and/or microbial contamination can lead to an unsatisfactory product. A buttery flavour note is produced by *Leuconostoc mesenteroides* subsp. *cremoris*. This bacterium converts acetaldehyde to diacetyl, thus reducing the "green" or yogurt-like flavour. A diacetyl to acetaldehyde ratio of 4:1 is desirable, whereas the green flavour is present when the ratio is 3:1 or less.

Proteolysis by the lactococci is necessary to afford growth of the *Leuconostoc* culture, and citrate is needed as substrate for diacetyl production Although cooking of the curd destroys virtually all bacteria capable of spoiling cottage cheese, washing and handling of the curd after cooking can introduce substantial numbers of spoilage microorganisms. It is desirable to acidify alkaline waters for washing cottage cheese curd to prevent solubilization of surfaces of the curd.

However, more pseudomonads can be adsorbed onto cottage cheese curd from wash water when adjusted to pH5 (40–45%) rather than adjusted to pH7 (20–30%). Flushing packages of cottage cheese or sour cream with CO_2 or N_2 suppressed the growth of psychrotrophic bacteria, yeasts, and molds for up to 112 days, but a slight bitterness can occur in cottage cheese after 73 days of storage. Cheesemakers can use the addition of high numbers of lactic acid bacteria to raw milk during storage to reduce the rate of growth of psychrotrophic microbes.

For fresh, raw milk, brined cheeses, gassing defects can be reduced by presalting the curd prior to brining and reducing the brine temperature to <12°C. Pasteurization will eliminate the risk from most psychrotrophic microbes, coliforms, leuconostocs, and many lactobacilli, so cheeses made from pasteurized milk have a low risk of gassiness produced by these microorganisms.

Most bacterial cells, including spores, can be removed from milk by centrifugation at about $9,000g$. The process, known as bactofugation, removes about 3% of the milk, called bactofugate. Kosikowski and Mistry invented and patented a process for recovering this bactofugate which is heated at 135°C for 3–4 s, then added back to the cheese milk. The process can reduce the population of butyric acid-producing spores by 98%. Spore-forming bacterial growth and subsequent gas production in aged, ripened cheeses can be minimized with a salt to moisture content of ≥3.0%.

Other potential inhibitors of butyric acid fermentation and gas production in cheese are the addition of nitrate, addition of lysozyme cold storage of cheese prior to ripening, direct salt addition to the cheese curd, addition of hydrogen peroxide, or use of starter cultures that form nisin or other antimicrobials. The most popular mold inhibitors used on cheeses are sorbates and natamycin. Sorbates tend to diffuse into the cheese, thereby modifying flavour and decreasing their concentration, whereas very little natamycin diffuses.

Electron beam irradiation, studied by Blank, Shamsuzzaman, and Sohal for mold decontamination of Cheddar cheese, can reduce initial populations of *Aspergillus ochraceus* and *Penicillium cyclopium* by 90% with average doses of 0.21 and 0.42 kGy, respectively. Since nearly all mold spores are killed by pasteurization practices that limit recontamination and growth, although difficult, are vital in prevention of moldy cheeses.

Modified atmosphere packaging (MAP) of cheeses can retard or prevent the growth of molds, and optimum MAP conditions for different types of cheeses were described by Nielsen and Haasum. For processed cheeses containing no active lactic acid starter bacteria, low O2 and high

CO2 atmospheres were optimum; for cheeses containing active starter cultures, atmospheres containing low O2 and controlled CO2 using a permeable film provided the best results. For mold-ripened cheeses requiring the activity of the fungi to maintain good quality, normal O2 and high, but controlled, CO2 atmospheres were best.

In Italian soft cheeses such as Stracchino, vacuum packaging decreased the growth of yeasts, resulting in a shelf life extension of >28 days. Processing times and temperatures used in the manufacture of cream cheese and pasteurized process cheese are able to eliminate most spoilage microorganisms from these products.

However, the benefit of the presence of competitive microflora is also lost. It is very important to limit the potential for recontamination, as products that do not contain antimycotics can readily support the growth of yeasts and molds. Sorbates can be added; however, their use in cream cheese is limited to amounts that will not affect the delicate flavour.

Prevention of Spoilage in Other Dairy Products

The high salt concentration in the serum-in-lipid emulsion of butter limits the growth of contaminating bacteria to the small amount of nutrients trapped within the droplets that contain the microbes. However, psychrotrophic bacteria can grow and produce lipases in refrigerated salted butter if the moisture and salt are not evenly distributed. When used in the bulk form, concentrated (condensed) milk must be kept refrigerated until used.

It can be preserved by addition of about 44% sucrose and/or glucose to lower the water activity below that at which viable spores will germinate (a_w 0.95). Lactose, which constitutes about 53% of the nonfat milk solids, contributes to the lowered water activity. When canned as evaporated milk or sweetened condensed milk, these products are commercially sterilized in the cans, and spoilage seldom occurs. Microbial growth and enzyme activity are prevented by freezing. Therefore, microbial degradation of frozen desserts occurs only in the ingredients used or in the mixes prior to freezing.

DRINK OF MILK PRODUCT ITEMS

There is but one real beverage and that is water. The other so-called beverages are foods, stimulants or sedatives. Milk is a rich food, one glass having as much food value as two eggs. Coffee, tea, chocolate and cocoa are stimulants, with sedative after-effects. Their food value depends largely on the amount of milk, cream and sugar put into them. Chocolate and cocoa are both drugs and foods. Alcohol is a stimulant at first, afterwards a sedative, and at all times an anaesthetic. When we think of drinking for the sake of supplying the bodily need of fluid, we should think of water and nothing else. If other liquids are taken, they should be taken as foods or drugs. Water is the best solvent known.

The alchemists of old spent much time and energy trying to find the universal solvent, believing that thereafter it would be easy to discover a method of making base metals noble. But they never found anything better than water. Water is the compound that in its various forms does most to change the earth upon which we live, and it is more necessary for the continuation of life than anything else except air. Pure water does not exist in nature, that is, we have never found a compound of the composition H_2O. Water always contains other matter.

The various salts are dissolved in it and it absorbs gases. The nearest we come to pure water is distilled. Pure water is an unsatisfied compound, and as soon as it is exposed it begins to absorb gases and take up salts and organic matter. Pure water differs from clean water. Clean or potable water is a compound which contains a moderate amount of salts, but very little of organic matter. Bacteria should be practically absent. Water that contains much of nitrogenous substances is unfit to use.

If the water is very hard, heavily loaded with salts, it should not be used extensively as a drink, for if too much of earthy and mineral matter is taken into the system, the body is unable to get rid of all of them. The result is a tendency for deposits to form in the body. In places where the water is excessively charged with lime it has been noticed that the bones harden too early, which prevents full development of the body.

If the bones of the skull are involved, it means that there will not be room enough for the brain. Such diseases are rare in this country, but in parts of Europe they are not uncommon. If the water is very hard, a good plan is to distill it and then add a little of the hard water to the distilled water. People who partake of an excessive amount of various salts can perhaps drink distilled water to advantage, but those who take but a normal amount of the salts in their foods should have natural water. Water forms three-fourths of the human body, more or less. It is needed in every process that goes on within the body. "To be dry is to die."

Water keeps the various vital fluids in solution so that they can perform their function. Without water there would be no sense of taste, no digestion, no absorption of food, no excretion of debris, and hence no life. The water is the vehicle through which the nutritive elements are distributed to the billions of cells of the body, and it is also the vehicle which carries the waste to the various excretory organs.

We can live several weeks without food, but only a few days without water. Hot water and ice-cold water are both irritants. Water may be taken either warm or cool. It is best to avoid the extremes. The amount of water needed each twenty-four hours varies according to circumstances. Two quarts is a favourite prescription. Those who eat freely of succulent fruits and vegetables do not need as much as those who live more on dry foods. Salt in excess calls for an abnormal amount of water, for salt is a diuretic, robbing the tissues of their fluids and consequently more water has to be taken to keep up the equilibrium.

Naturally, more water is required when the weather is hot than when it is cool. On hot days warm water is more satisfying and quenches thirst more quickly than ice water. Warm water also stimulates kidney action, which is often sluggish in summer. Ice water is the least satisfactory of all, for the more one drinks the more he wants. A normal body calls for what water it needs, and no more. An abnormal body is no guide for either the amount of food or drink necessary. Many people do not like the taste of water, especially in the morning.

This means that the body is diseased. To a normal person cool water is always agreeable when it is needed, and it is needed in the morning. People with natural taste do not care for ice water, but other water is relished. The common habit of drinking with meals is a mistake. Man is the only animal that does this, and he has to pay dearly for such errors. Taking a bite of food and washing it down with fluid lead to undermastication and overeating, and then the body suffers from autointoxication. A mouthful of food followed by a swallow of liquid forces the contents of the mouth into the stomach before the saliva has the opportunity to act.

The best way is to drink one or two glasses of water in the morning before breakfast. Partake of the breakfast, and all other meals, without taking any liquid. Sometimes there is a desire for a drink immediately after the meal is finished. If so, take some water slowly. If it is taken slowly a little will satisfy. If it is gulped down it may be necessary to take one or two glasses of water before being satisfied.

Those who have a tendency to drink too much during warm weather will find very slow drinking helpful in correcting it. If there is any digestive weakness, the liquid taken immediately after a meal should be warm and should not exceed a cupful. Those with robust digestion may take cool water.

Cold water chills the stomach. Digestion will not take place until the stomach has reached the temperature of about one hundred degrees Fahrenheit again, and if the stomach contents are chilled repeatedly the tendency is strong for the food to ferment pathologically, instead of being properly digested. For this reason it is not well to drink while there is anything left in the stomach to digest.

As stomach digestion generally takes two or three hours at least, it is well to wait this long before taking water after finishing a meal, and then drink all that is desired until within thirty minutes of taking the next meal. If the thirst should become very insistent before two or three hours have elapsed since eating, take warm water. Those who eat food simply prepared and moderately seasoned are not troubled much with excessive thirst. Two quarts of water daily should be sufficient for the adults under ordinary conditions. Here, as in eating, no exact amount will fit everybody. Make a habit of drinking at least a glass of water before breakfast, cleaning the teeth and rinsing the mouth before swallowing any, and then take what water the body asks for during the rest of the day. Taking too much water is not as injurious as overeating, but waterlogging the body has a weakening effect.

To drink with the meals is customary, not because it is necessary, but because we have a number of drinks which appeal to many people. Water is the drink par excellence. A food-beverage that is used by many is cambric tea, which is made of hot water, one-third or one-fourth of milk and a little sweetening. Children generally like this on account of the sweetness. It may be taken with any meal, when fluid is needed, but the amount should be limited to a cupful. It is not well to dilute the digestive juices too much.

The water taken in the morning helps to start the body to cleanse itself. Water drinking is a great aid in overcoming constipation. Constipated people generally overeat. Less food and more water will prove helpful in overcoming the condition. Unfortunately for the race, we have accustomed ourselves to partake of beverages containing injurious, poisonous substances. Inasmuch as this is the place to discuss the drugs contained in coffee and tea, we shall take the liberty of dwelling upon other habit-forming substances.

They are all a part of the drug addictions of the race. For scientific discussion of these various substances we refer you to technical works. Coffee, tea and chocolate contain a poisonous alkaloid which is generally called caffeine. The theine in tea and the theobromine in cocoa are so similar to caffeine that chemists can not differentiate them. These drinks when first taken cause a gentle stimulation under which more work can be done than ordinarily, but this is followed by a reaction, and then the powers of body and mind wane so much that the average output of work is less than when the body is not stimulated.

The temporary apparently beneficial effect is more than offset by the reaction and therefore partaking of these beverages makes people inefficient. Coffee is very hard on the nerves, causing irritation, which is always followed by premature physical degeneration. Experiments of late indicate that children who use coffee do not come up to the physical and mental standard of those who abstain. The effect on the adults is not so marked because adults are more stable than children.

Those who are not used to coffee will be unable to sleep for several hours after partaking of a cup. Some people drink so much of it that they become accustomed to it. Coffee is not generally looked upon as one of the habit-forming drugs, but it is. However, of all the drugs which create a craving in the system for a repetition of the dose, coffee makes the lightest fetters.

It is surprising how often health-seekers inform the adviser that they "can not get along without coffee". If they would take a cup a few times a year, it would do no harm, but the daily use is harmful to all, even if they feel no bad effects and make it "very weak," which is a favourite statement of the women. Smoking, drinking beer and drinking coffee have a tendency to overcome constipation in those who are not accustomed to these things, but their action can not be depended upon for any length of time and the cure is worse than the disease. Tea drinking has much the same effect as coffee drinking, except that it is decidedly constipating. Perhaps this is because there is considerable of the astringent tannin in the tea leaves.

Chocolate is a valuable food. Those who eat of other aliments in moderation may partake of chocolate without harm, but if chocolate is used in addition to an excess of other food, the results are bad. The chocolate is so rich that it soon overburdens some of the organs of digestion, especially the liver. The Swiss consume much of this food and it is valuable in cases where it is necessary to carry concentrated rations.

Alcohol in some form seems to have been consumed by even very primitive people as far back as history goes. The Bible records an early case of intoxication from wine, and beer was brewed by the ancient Egyptians.

So much has been consumed that some people have a subconscious craving for it. There are cases on record where the very first drink caused an uncontrollable demand for the drug. Fortunately these cases are very rare. Alcohol is really not a stimulant, though it gives a feeling of glow, warmth and well-being at first, but this is followed by a great lowering of physical power, which gives rise to disagreeable sensations. Then the drinker needs more alcohol to stimulate him again. Then there is another depression with renewed demand: There is no end to the craving for the drug once it has mastered the individual. The lungs, heart, digestive organs, muscles, in fact, every structure in the body loses working capacity. Alcohol seems to have a special affinity for nervous tissue. A glass of beer or wine taken daily is no more harmful than a cup of coffee per day, but the coffee drinker does not make of himself such a public nuisance and menace as the man often does who drinks alcohol to excess. Formerly, it was respectable to drink. Some of our most noted public men were drunkards. Now a drunkard could not maintain himself in a prominent public position very long. To drink like a gentleman was no disgrace. Now real gentlemen do not get drunk.

In backward Russia they are becoming alarmed about the inroads of vodka, and are trying to decrease its consumption. France is trying to teach total abstinence to its young men because it disqualifies so many of them from military service to drink. Scandinavia is temperance territory. The German Kaiser has recently given a warning against drinking. The United States discourages drinking in the army and navy. Field armies are not supplied with alcoholics. Drinking is becoming disreputable.

It is very difficult to prove the harm done by excessive drinking of tea and coffee, also by the use of much tobacco, even if we do know that it is so. Everyone knows something about the `deleterious` effect of alcohol upon the consumer. Solomon wrote: "Wine is a mocker, strong drink is raging, and whosoever is deceived thereby is not wise. Who hath wounds without cause? Who hath redness of eyes?" Alcohol permanently impairs both body and mind. Depending on how much is taken, it may cause various ills, ranging from inflammation of the stomach to insanity. It reduces the power of the mind to concentrate and it diminishes the ability of the muscles to work. It reduces the resistance of the body and shortens life. Its first effect is to lull the higher faculties to sleep. Most drunkards do not recover from their disease, for drunkenness is a disease. The various drugs given to cure the afflictions are delusions. Strengthening the body, mind and the will and instilling higher ideals are the best methods of cure. Suggestive therapeutics, and the awakening of a strong resolve for a better life are powerful aids. Proper feeding should not be overlooked, for bad habits do not flourish in a healthy body.

Civilization necessitates self-control and considerable self-denial. Those who go in the line of least resistance are on the road to destruction. It is often necessary to overcome habits which produce temporary gratification of the senses. Warden Tynan of the Colorado Penitentiary, 96 per cent of the prisoners are brought there because they use alcohol. It is also well known that moral lapses are most common when the will is weakened through the use of liquor.

Those who have the welfare of the race at heart are therefore compelled to give considerable thought to this subject. Past experience, it will not help to try to legislate sobriety into the people. Education and industrialism are the factors which it seems to me will be most potent in solving the alcohol problem. Morality, which in the last analysis is a form of selfishness, will teach many that it is poor policy to reduce one's efficiency and thereby reduce the earning capacity and enjoyment of life.

More and more the employers of labour will realise that the use of alcohol decreases the reliability and worth of the worker. Many will take steps like the following: "In formal recognition of the fact, established beyond dispute by the tests of the new psychology, that industrial efficiency decreases with indulgence in alcohol and is increased by abstinence from it, the managers of a manufacturing establishment in Chester, Penn., have attacked the temperance problem from a new angle. "Unlike many railways and some other corporations, they do not forbid their employees to drink, but they offer 10 per cent advance in wages to all who will take and keep—the teetotaler's pledge. Incidentally, a breaking of the promise will mean a permanent severance of relations, but there is no emphasizing of that point, it being confidently expected that the advantage of perfect sobriety will be as well realised on one side as on the other."

Business has during the past two centuries been the great civilizer, the great moral teacher. It has found that honesty and righteousness pay and that injustice is folly. Business has led the way to the acceptance of a new ethics, and new morals. What has been said about alcohol applies to tobacco in a much smaller degree. The use of tobacco seems to lead to the use of alcohol. It retards the development of children. It is surely one of the causes of various diseases. Tobacco heart, sore throat and indigestion are well known to physicians. Tobacco contains one of the deadliest of poisons known. One-sixteenth of a grain of nicotine may prove fatal. The reason there are so few deaths from acute tobacco poisoning is that but very little of the nicotine is absorbed. Men who chew tobacco make themselves disagreeable to others. Smoking of cigarettes is to be condemned not only because it poisons the body, but causes inattention and inability to concentrate on the part of the smoker, as well.

Every little while he feels the desire to take a smoke, and if smoking is forbidden he devises means of getting away. He robs his employer of time for which he is paid and injures himself. The ability to work is decreased by indulgence in smoking. Recent experiments show that for a short time there is increased activity after a smoke, but the following depression is greater than the stimulation, so there is an actual loss.

A few years ago, according to Mr. Wilson, who was then Secretary of Agriculture, there were about 4,000,000 drug addicts or "dope fiends" in the United States. Without doubt this estimate was too high, for the proportion of addicts in the country is not as great as in the large cities. The drugs chiefly used are cocaine, opium, laudanum, morphine and heroin. These drugs are much more destructive than alcohol. Cocaine and heroin are the worst. It is very difficult to stop using any of them once the habit has been formed. Nearly every "fiend" dies directly or indirectly from the effect of his particular drug. Every one weakens the body so that there is not much resistance to offer to acute diseases. Every one destroys the will power so that a cure is exceedingly difficult.

It is well to bear in mind that all are not possessed of strong enough will power to resist their cravings and that some take to cocaine when they can not get liquor. Cocaine is far worse than alcohol.

People should be very careful about taking patent medicines. There is no excuse for taking them. The most popular ones have as their basis one of the habit-forming drugs. Most of the soothing syrups contain opium in some form. To give babies opiates is a grave error, to speak mildly. It weakens the child, may lay the foundation for a deadly habit later in life, and often an overdose kills outright. Well informed mothers avoid such drugs and keep their children reasonably quiet by means of proper care.

Many of the remedies for nasal catarrh and hay fever contain much cocaine. Cocaine is an astringent and a painkiller and people mistake the temporary lessening of discharge from the nose and disappearance of pain for curative effects. But there is nothing curative about it. In a short time the mucous membrane relaxes again and then the discharge is re-established. The nerves which were put out of commission resume their function and then the pain reappears.

Opium or one of its derivatives is generally present in the patent medicines given for coughs. Opium is also an astringent and will suppress secretions, but this is not a cure. Excessive secretions are an indication that the body is surcharged with poison and food. Let them escape and then live so that there will be internal cleanliness and then there will be no more coughs and colds.

The unfortunate people who get into the habit of using these drugs degenerate physically, mentally and morally. They need more and more of their drug to produce the desired effect until they at last take enough daily to kill several normal men. Sometimes they are able to keep everybody in ignorance of what they are doing for years. They develop slyness and secretiveness.

They become very suspicious. They are nearly always untruthful, and those who deal with them are surprised and wonder why those who used to be open and above-board now are furtive and dishonest. They often lie when there is not the slightest excuse for it. The moral disintegration is often the first sign noticed.

After habitually using any of these drugs for a while the body demands the continuation and if the victim is deprived of his accustomed portion there will be a collapse with intense suffering. Every tortured nerve in the body seems to call out for the drug. The victim will do anything to get his drug. He will lie, steal, and he may even attack those who are caring for him. For the time being he is insane. Many professional men use cocaine. It is a favourite with writers. It often shows in their work. Those who write under the inspiration of this drug often do some good work, but they are unable to keep to their subject. Their writings lack order. We have enough of such writings to have them classified as "cocaine literature". If there are 4,000,000, or even fewer, of these people in our land, it is a serious problem, for every one is a degenerate, to a certain degree. If the medical profession and the druggists would cooperate it would be easy enough to prevent the growth of a new crop of dope fiends.

Of course, people would have to stop taking patent medicines, which often start the victims on the road to degeneration.

Then the physicians should stop prescribing habit-forming drugs, as well as all other drugs, and teach the people that physical, mental and moral salvation come through right living and right thinking. Unfortunately the medical profession is careless and is responsible for the existence of many of the drug addicts. A patient has a severe pain. What is the easiest way to satisfy him? To give a hypodermic injection of some opiate. The patient, not realizing the danger, demands a pain-killer every time he suffers. He soon learns what he is getting and then he goes to the drug store and outfits himself with a hypodermic outfit and drugs, and the first thing he knows he is a slave, in bondage for life.

This is no exaggeration. There are hundreds of thousands of victims to the drug habit who trace their downfall to the treatment received at the hands of reputable physicians, who do not look upon their practice with the horror it should inspire because it is so common. Doctors do not always bury their mistakes. Some of them walk about for years.

In spite of laws against the sale of various drugs, they can be obtained. There are doctors and druggists of easy conscience who are very accommodating, for a price. There is no legitimate need for the use of one-hundredth of the amount of these drugs that is now consumed. A local injection of cocaine for a minor operation is justifiable, but none of the habit-forming drugs should be used in ordinary practice to kill pain, for the proper application of water in conjunction with right living will do it better and there are no evil after effects. Massage is often sufficient.

To show a little more clearly how some people become addicted to drugs, let us consider one of the latest, heroin: A few years ago this drug, which is an opium derivative, was practically unknown. It is much stronger than morphine and consequently the effect can be obtained more quickly by means of a smaller dose. Physicians thought at first that it was not a habit-forming drug, for they could use it over a longer period of time than they could employ morphine, without establishing the craving and the habit.

So they began to prescribe heroin instead of morphine, and many a morphine addict was advised to substitute heroin. All went well for a short while, until the victims found that they were enslaved by a drug that was even worse than morphine. Now, thanks chiefly to the medical profession, it is estimated that we have in our land several hundred thousand heroin addicts. Sallow of face, gaunt of figure, looking upon the world through pin-point pupils, with all of life's beauty, hope and joy gone, they are marching to premature death.

The medical profession furnishes more than its proportion of drug addicts. They know the danger of the drugs, but familiarity breeds contempt. If the public but knew how many of their medical advisers, who should always be clear-minded, are befuddled by drugs, there would be a great awakening. One eminent physician who has now been in practice about forty-five years and has had much experience with drug addicts, has said that according to his observations, about one physician in four contracts the drug habit. I believe this is exaggerated, but I am acquainted with a number of physicians who are addicts.

Physicians who smoke do not condemn the practice. Those who drink are likely to prescribe beer and wine for their patients. Those who are addicted to drugs use them too liberally in their practice. Those who have watched the effects of the various drugs, from coffee to heroin, must condemn their use. It is true that an occasional cup of coffee or tea, a glass of wine or beer does no harm. A cigarette a week would not hurt a boy, nor would on occasional cigar harm a man.

But how many people are willing to indulge occasionally? The rule is that they indulge not only daily, but several times a day, and the results are bad. One bad habit leads to another, and the time always comes when it is a choice between disease and early death on one hand, and the giving up of the bad habits on the other, and when this time comes the bonds of habits are often so strong that the victim is unable to break them. We realise that knowledge will not always keep people out of temptation and that some individuals will take the broad way that leads to destruction in spite of anything that may be said. Youth is impatient of restraint and ever anxious for new experiences.

Regarding this serious matter of destructive drug use, much could be done by teaching people their place in society: That is, what they owe to themselves, their families and the public in general. In other words, teach the young people the higher selfishness, part of which consists of considerable self-control, self-denial and self-respect.

Drugs are too easy to obtain today. Some day people will be so enlightened that they will not allow themselves to be medicated. This is the trend of the times. Until such a time comes, society should protect itself by making it very difficult to get any of the habit-forming drugs. If

Prevention of Spoilage in Milk

necessary, the free hand of the physician should be stayed. Much of the confidence blindly given him is misplaced.

HYGIENE DURING MILKING

There are several possible causes of contamination during milking. In a normal, healthy cow very low numbers of bacteria are found inside the udder and the teats. Cows possess various mechanisms to prevent the entry of bacteria. To avoid problems while milking, it is important that an animal become accustomed to the activity. It will then know that it will be milked, and will react positively to it. Such positive behaviour can start if, for instance, it hears milk cans clanging, feels its udder being cleaned, etc. Then the animal is easier to milk and gives more milk.

Stress and unrest make the cows move too much and kick; consequently more dirt and manure can enter into the milk. When a cow has an udder infection (mastitis), its milk will be contaminated with the bacteria that cause the udder infection, and that may produce pus and sometimes blood. Milk from these animals should not be used in any way. Mastitis can be prevented by maintaining good hygiene and avoiding injury to the teats during milking.

An infected udder is not always easy to see. When an udder infection occurs, it is advisable to remove milk from the udder very frequently (*e.g.* every 3 hours by hand). The number of microorganisms in the udder is thus reduced. Be aware, however, that milking an infected udder by machine or by hand is often painful for the animal. The animal will kick frequently and this can be an important source of contamination of healthy cows. Bacteria can be transferred from the skin or teats to the milk, even with healthy dairy cattle.

It is therefore important to clean the udder before milking. Wipe the udder clean with a dry, clean, preferably disposable cloth to prevent infection. If the teats or udder are really dirty, they must first be washed with clean, hand-warm water and a clean cloth and then dried with a clean towel. Cleaning the udder improves the cleanliness of the milk and makes milking easier. Skin and hair can also be sources of infection. Do not feed animals first before milking, it may create a lot of dust. The floor is clean, and be careful when clearing dung, mud or dust.

A clean, well-illuminated milking place and fresh surrounding air are essential to maintaining good hygiene. Insects such as flies and cockroaches can also be sources of infection. Try to control them as they can carry many bacteria and viruses. When milking, the milk is caught in a pail or bucket. Dirty milking equipment is the main source of infection of milk. If residues of milk remain in the equipment because of improper cleaning and drying, bacteria will develop in these residues. These bacteria are already accustomed to the milk and will multiply rather quickly during transport and storage of milk in the equipment.

Use pails and buckets that are smooth on the inside, for instance seamless metal buckets. All milking equipment should be thoroughly cleaned immediately after each use. Use soap or other detergents if necessary. Make sure that the water used is clean. If you are in doubt, boil it for several minutes or add chlorine. Very important: after cleaning, the equipment should be stored upside down in such a way that the inside of the buckets and cans dry. This prevents the remaining bacteria from growing.

The person milking plays the most important part in maintaining proper hygiene during production. He or she keeps an eye on the condition of the animal, chooses the milking place and cleans all the equipment. He or she should have clean hands and wear clean clothes. If the milker suffers from tuberculosis, salmonella infection, dysentery or some other disease, the risk of contamination of the milk becomes very high; it would be wise to have somebody else take over. This is also the case if the milker has open wounds or ulcers.

Hygiene during Storage and Processing

By now you should know that milk should be processed as quickly as possible after milking and that it should be properly stored in order to minimise its chances of spoiling. It is best to filter fresh milk through a filter or clean cloth. This will remove visible dirt that might have entered into the milk. Clean or replace the cloth during filtering or filter the milk several times. The cloth should be thoroughly cleaned after use and then left to dry in the sun.

In tropical conditions, raw milk, *i.e.* non-pasteurised milk, goes off within a few hours. It must therefore be kept cool and quickly pasteurised and again cooled to a temperature of 4°C if possible. Properly pasteurised and cooled milk can be kept for a few days, even in a warm climate. If you are not able to cool milk below 10°C, then do not mix different batches.

Even if the older milk is still good, you will end up with an increase in bacterial growth and reduction of the overall quality. Use clean equipment for storage. Containers that are clear, such as glass, should

Prevention of Spoilage in Milk

be stored in the dark as light reduces the quality of milk. Clean your equipment with clean water.

Cleaning and Disinfection

Utensils must be cleaned in such a way that all dirt, food residues, feed and micro-organisms are removed from the surface of the equipment. Dirty saucepans, jugs, milking equipment and utensils should be cleaned immediately after use. Washing soda (sodium carbonate) dissolved in hot water is an excellent cleaning agent. It may be useful to disinfect equipment in order to kill any remaining harmful micro-organisms. You can use a chloride solution such as bleach (sodium hypochlorite).

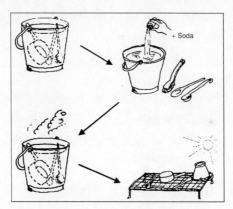

Fig. Cleaning Utensils

A proper way of cleaning your equipment is the following:
- Start cleaning immediately after milking, so that milk residues will not dry and stick on the buckets and utensils.
- Rinse well with water.
- Scrub the tools in a hot soda solution (1.5 tablespoons of soda to 5 litres of water), using a small amount of water to dissolve the soda before adding it to the rest of the water.
- Rinse well with hot water.
- Buckets, tubs, etc., should be turned upside down on a rack during storage; the water can then drain and no dirt or dust can enter. Let the utensils dry to prevent bacterial growth.

Well-cleaned tools are nearly sterile, only a small part of the bacteria remains on the tools. If these tools dry during storage hardly any bacteria will be present. In that case disinfection is not necessary. Tools which are used for storage of pasteurised milk or for cheese making and which do not get a heat treatment together with the milk can be disinfected after cleaning or before use.

Proceed as follows:
- Clean all your equipment properly. The following step will be ineffective if the utensils are not clean to start with.
- Disinfect in a chloride or bleach solution after cleaning or shortly before use. Add 2 tablespoons of bleach per 4. 5 litres of water.

It is advisable to use stainless steel equipment, cheesecloth and wooden utensils. Tools or any other equipment made from aluminium should not be washed in a strong soda solution, as soda attacks aluminium. Iron utensils will rust in a strong chloride solution. Therefore rinse and dry these utensils immediately after cleaning and disinfection.

If you have no cleaning agents-like soda-or disinfectants, you can disinfect your equipment as follows:
- Thoroughly clean the utensils using clean water.
- Rinse with a soap solution.
- Dry the equipment on a rack in the sun upside down or rinse with boiling water.

PASTEURISED MILK PRODUCTS

Salmonella

Salmonellae are not able to survive the typical minimum pasteurisation processes generally prescribed in legislation. Therefore, their presence indicates that the process has not been carried out effectively, or that post-process contamination has occurred. For example, an outbreak of salmonellosis in Kentucky in 1984 was associated with pasteurised milk, but an investigation of the dairy concerned showed that pasteurisation temperatures were inadequate, and could have been as low as 54.5°C for 30 minutes.

An outbreak caused by *Salmonella braenderup* in the UK in 1986 was also associated with pasteurised milk, and on this occasion the pasteuriser was found to be poorly designed and incorrectly operated, probably resulting in the application of an inadequate heat treatment. In 1985, one of the largest outbreaks of salmonellosis in US history occurred in Illinois. Almost 200,000 people were affected, and were associated with pasteurised low-fat 2% milk contaminated with *S. typhimurium*. Investigations at the dairy

Prevention of Spoilage in Milk

plant involved revealed no evidence of inadequate pasteurisation, and the outbreak strain was not found to be abnormally heat resistant.

Although the cause of the outbreak has never been completely explained, the investigation did discover a possible cross-connection between raw and pasteurised milk, which may have been the source of contamination. In 1998, an outbreak of salmonellosis in Lancashire, caused by a multiresistant strain of *S. typhimurium* DT104, affected 86 people. This outbreak was also linked to defective pasteurisation of milk at a dairy on a local farm.

Consumption of raw milk or raw milk products have been responsible for 62 and 29 cases of diarrheal illness caused by *S. typhimurium* in 2003 and 2007, respectively, in the states of Ohio and Pennsylvania. Since salmonellae are occasional contaminants of raw milk, they may sometimes enter the processing environment. It is very important that contamination of the post-pasteurisation plant is not allowed to occur and effective precautions and monitoring procedures, based on HACCP principles, are necessary to prevent this.

Campylobacter Spp.

Campylobacter spp. are not capable of surviving milk pasteurisation treatments, and cannot grow in raw or pasteurised milk, although they are able to survive for long periods in milk at refrigeration temperatures. Nonetheless, outbreaks of campylo-bacteriosis associated with pasteurised milk have occurred. For example, a large outbreak in the UK in 1979 caused by *Campylobacter jejuni* was estimated to have affected at least 2,500 schoolchildren, and was associated with free milk provided in schools. Although conclusive evidence was absent, it seems likely that raw milk may have bypassed the pasteurisation process.

A more recent outbreak in 2001 involved 75 people and was linked to the consumption of unpasteurised milk procured thorough a cow leasing programme. Birds are known to be an important reservoir of *Campylobacter* infection, and the tendency of some birds to peck through the foil tops of doorstep-delivered milk bottles is becoming recognised as an important source of infection in parts of the UK.

Some individual cases have been attributed to this cause, and, in one instance in 1990, the organism was isolated from the beaks of jackdaws and magpies as well as the contaminated milk. More recently, an outbreak thought to be associated with bird-pecked milk was reported.

Listeria Monocytogenes

There has been some discussion regarding the potential for *L. monocytogenes* in milk to survive pasteurisation. An outbreak of listeriosis in Massachusetts during 1983 resulted in 49 cases, 14 of whom subsequently died. Epidemiological evidence strongly suggested an association with consumption of pasteurised whole and low-fat milk, although this could not be confirmed microbiologically. The investigation failed to reveal any evidence of inadequate pasteurisation and the organism could not be found in environmental samples in the dairy, suggesting that postprocess contamination was unlikely.

However, samples of raw milk taken from farms supplying the plant were found to be positive for *L. monocytogenes* serotype 4b, and the investigators concluded that survival of some organisms through pasteurisation was the most likely cause of the outbreak. Three deaths and a miscarriage in Boston, USA between 2007-8 have been linked to presence of *Listeria* in pasteurised milk. So far, investigations have found nothing wrong with its pasteurisation process.

Furthermore, in a survey of pasteurised milk conducted in Spain, *L. monocytogenes* was recovered from six out of 28 samples heated at 78°C for 15 seconds. The explanation offered for both these findings was that the organisms might have been protected during heat treatment within leucocytes in the milk. However, this effect has not been conclusively demonstrated, and *L. monocytogenes* has not yet been shown to have survived pasteurisation in milk subjected to minimum HTST pasteurisation requirements of 71.7°C for 15 seconds.

For these reasons, it is currently accepted that existing pasteurisation processes are adequate to inactivate the organism in milk. *L. monocytogenes* is likely to be present in wet dairy processing environments, and post-process contamination is therefore a particular hazard. The organism has been shown to be capable of significantly more rapid growth in pasteurised milk than in raw milk at 7°C, and is also capable of growth at 4°C in pasteurised milk. Therefore, effective HACCP-based controls to prevent post-process contamination are critical, particularly the cleaning and sanitising of all milkcontact surfaces. Adequate temperature control is also important.

Verotoxigenic Escherichia Coli

Dairy cattle are an important reservoir for *E. coli* O157: H7 and this organism may therefore be present in raw milk, usually through faecal contamination. For this reason, raw milk is a high-risk food for this serious intestinal pathogen, and there have been a number of small outbreaks of infection associated with its consumption. However, *E. coli* O157: H7 is not a heat-resistant organism and there is no evidence that it is able to survive pasteurisation. Despite this, there have been outbreaks associated with pasteurised milk.

In 1994, an outbreak in Scotland affected over 100 people and was associated with consumption of pasteurised milk from a local dairy. The outbreak strain was eventually recovered from cows on one of the farms supplying the dairy, from a bulk milk tanker, and from a pipe transferring milk from the pasteuriser to the bottling machine. Whether this outbreak was the result of faulty pasteurisation or post-process contamination was unclear, but, in either case, the raw milk is likely to have been the original source of the organism.

In 1999, a serious outbreak occurred in Cumbria in the north-west of England, which was also associated with pasteurised milk from a local dairy. There were at least 60 confirmed cases involved, and the cause was thought to be a fault in the operation of the pasteuriser. The first general outbreak of verocytotoxin-producing *E. coli* in Denmark occurred in 2004 and involved 25 patients; 18 children and seven adults. It was thought to be due to the consumption of a particular kind of organic milk from a small dairy.

Environmental and microbiological investigations at the suspected dairy did not confirm the presence of the outbreak strain, but the outbreak stopped once the dairy was closed and thoroughly cleaned. *E. coli* O157 is not reported to be able to grow in raw or pasteurised milk stored at 5°C, but may grow slowly at higher temperatures. However, since the infective dose of this pathogen is thought to be very low, effective pasteurisation and the prevention of post-process contamination are critical to ensure product safety.

Yersinia Enterocolitica

Although there has been a question about the ability of *Y. enterocolitica* to survive milk pasteurisation, the majority of the evidence indicates that it is inactivated. Three different strains of *Y. enterocolitica* were reported to have D-values of 0.24-0.96 minutes at 62.8°C. Therefore, the presence of the organism in pasteurised milk is likely to be the result of post-process contamination. There have been several *Y. enterocolitica* outbreaks associated with pasteurised milk.

In 1976, an outbreak affecting 36 children was associated with the consumption of contaminated chocolate milk. It was thought that the organism was introduced to the product during mixing of chocolate syrup with pasteurised milk, without any subsequent heat process. Another outbreak in 1982 was the largest foodborne yersiniosis outbreak ever recorded in the USA, and was also associated with pasteurised milk.

It is thought that several thousand people may have developed illness, although the organism was not isolated from milk or environmental samples at the dairy. It was found that surplus milk was used to feed pigs and that the crates used to transport this milk were stored on the ground at the farm and could have become contaminated with pig faeces.

Since pigs are a well known reservoir for *Y. enterocolitica*, it was thought that inadequate washing of the crates allowed the organism to survive in mud on them, and subsequently contaminate the external surfaces of milk cartons. *Y. enterocolitica* is capable of psychrotrophic growth, and could therefore multiply in pasteurised milk during storage. Measures should therefore be taken to prevent post-process contamination as with *L. monocytogenes*.

Staphylococcus Aureus

Staph. aureus is only rarely involved in food poisoning associated with consumption of pasteurised milk, although enterotoxigenic strains can be found as contaminants in raw milk. This may be because *Staph. aureus* does not generally grow at temperatures below 7°C, and enterotoxin production is inhibited at low temperatures. The organism is also known to be inhibited by the presence of competing species. Nevertheless, an outbreak in California affecting 500 school children was associated with chocolate-flavoured milk. The cause was thought to be growth of *Staph. aureus* in raw milk, and the subsequent persistence of the heat-stable enterotoxin through pasteurisation.

In June and July 2000, a very large outbreak of staphylococcal food poisoning was reported in Japan, associated with consumption of pasteurised low fat milk. Over 14,500 people were said to have been affected. The outbreak was unusual in that the thermal processes had destroyed staphylococci in milk but *Staphylococcus* enterotoxin A had retained enough activity to cause intoxication. SEA exposed at least twice to pasteurisation at 130°C for 4 or 2s retained both immunological and biological activities, although it had been partially inactivated.

Bacillus Spp

Psychrotrophic *Bacillus* spp. present in raw milk may survive pasteurisation and then become dominant in the pasteurised milk, potentially causing spoilage. Concerns have been expressed that some psychrotrophic strains of *B. cereus* may be able to produce toxin in milk at refrigeration temperatures, but it seems likely that obvious spoilage would occur before sufficient toxin production had taken place to cause illness. Even so, *B. cereus* was isolated at levels of 4×10^5/g from pasteurised milk associated with 280 food poisoning cases in the Netherlands in 1989.

Mycobacterium Avium Subsp. Paratuberculosis

MAP is the causative organism of Johne's disease in cattle, a chronic wasting disease, and may occasionally be present in raw milk. Evidence linking MAP to a chronic inflammatory bowel condition in humans, called Crohn's disease, is becoming increasingly compelling. Concerns have been raised that MAP might be able to survive pasteurisation if present at levels above 100 cells per ml, especially if clumps of cells are present, and that pasteurised milk may therefore be a vehicle for Crohn's disease.

On the basis of new heat-resistance studies, many UK dairies have increased pasteurisation times to from 15 to 25 seconds. A survey of the level of contamination of pasteurised milk by MAP over a 17 month period, in 1999-2000, revealed a mean of 1.6% of raw and 1.8% of pasteurised samples were positive for MAP cultures indicating that commercially pasteurised milk may occasionally contain low levels of viable MAP. The potential public health impact of this situation is, however, still uncertain given that an association with Crohn's disease in humans remains unproven.

Viruses

A number of viruses have been shown to be present in raw milk, although many of these, such as Foot and Mouth Disease Virus (FMDV), are not pathogenic to humans. However, raw milk has been implicated in outbreaks of hepatitis and poliomyelitis. Some viruses, including poliovirus, are completely inactivated by pasteurisation, but this seems not to be the case with others, such as FMDV, if the virus is naturally present rather than inoculated. There is therefore the possibility that other viruses pathogenic to humans may survive at low levels, but, in bulk milk processing systems, it is thought unlikely that sufficient viruses will be present to infect consumers.

Toxins

Mycotoxins may be present in milk as a result of the ingestion of mouldy and contaminated feed by cattle. Feed contaminated by aflatoxin B1 as a result of the growth of *Aspergillus flavus* or *Aspergillus parasiticus* has been shown to give rise to the presence of aflatoxin M1 in the milk of dairy cows consuming it. However, only a small percentage of the ingested toxin appeared in the milk. Aflatoxins are persistent compounds and are not greatly affected by milk processing, and could therefore be present in pasteurised, packaged milk. However, recent surveys suggest that contamination of the milk supply is very limited and well within acceptable levels.

PATHOGENS: GROWTH AND SURVIVAL

Raw milk

Before the adoption of routine pasteurisation, milk was an important vehicle for the transmission of a wide range of diseases, including typhoid, brucellosis and diphtheria. Pasteurisation and improvements in veterinary medicine have seen a very large reduction in the incidence of such traditionally milkborne diseases.

However, raw milk may still contain a very wide range of pathogens, including *Salmonella* spp., *E. coli* O157, *L. monocytogenes* and *Campylobacter* spp. derived from the milk animals, the environment or from farm workers and milking equipment. Pathogens may be present even in hygienically produced milk of generally good microbiological quality. In short, raw milk is a potentially hazardous product, the microbiological safety of which cannot be assured without the use of pasteurisation or an equivalent process. Recent milk-associated outbreaks of infectious intestinal disease in the UK have been shown to be caused mainly by unpasteurised or inadequately pasteurised milk products.

FACTORS AFFECTING SPOILAGE

Spoilage of Fluid Milk Products

The shelf life of pasteurized milk can be affected by large numbers of somatic cells in raw milk. Increased somatic cell numbers are positively correlated with concentrations of plasmin, a heat-stable protease, and of lipoprotein lipase in freshly produced milk. Activities of these enzymes

Prevention of Spoilage in Milk

can supplement those of bacterial hydrolases, hence shortening the time to spoilage.

The major determinants of quantities of these enzymes in the milk supply are the initial cell numbers of psychrotrophic bacteria, their generation times, their abilities to produce specific enzymes, and the time and temperature at which the milk is stored before processing. Several conditions must exist for lipolyzed flavour to develop from residual lipases in processed dairy foods, that is, large numbers (>10^6 CFU/ml) of lipase producers, stability of the enzyme to the thermal process, long-term storage and favourable conditions of temperature, pH, and water activity.

Spoilage of Cheeses

Factors that determine the rates of spoilage of cheeses are water activity, pH, salt to moisture ratio, temperature, characteristics of the lactic starter culture, types and viability of contaminating microorganisms, and characteristics and quantities of residual enzymes. With so many variables to affect deteriorative reactions, it is no surprise that cheeses vary widely in spoilage characteristics. Soft or unripened cheeses, which generally have the highest pH values, along with the lowest salt to moisture ratios, spoil most quickly.

In contrast, aged, ripened cheeses retain their desirable eating qualities for long periods because of their comparatively low pH, low water activity, and low redox potential. For fresh, raw milk pasta filata cheeses, Melilli determined that low initial salt and higher brining temperature (18%C) allowed for greater growth of coliforms, which caused gas formation in the cheese. Factors affecting the growth of the spoilage microorganisms, *Enterobacter agglomerans* and *Pseudomonas* spp. in cottage cheese, were higher pH and storage temperature of the cheese.

Some of the spoilage microorganisms were able to grow at relatively low pH values when incubated at 7°C and were able to grow at pH 3. 6 when grown in media at 20°C. Rate of salt penetration into brined cheeses, types of starter cultures used, initial load of spores in the milk used for production, pH of the cheese, and ripening temperature affect the rate of butyric acid fermentation and gas production by *C. tyrobutyricum*.

Fungal growth in packaged cheeses was found to be most significantly affected by the concentration of CO_2 in the package and the water activity of the cheese. Cheddar cheese exhibiting yeast spoilage had a high moisture level (39.1%) and a low salt in the moisture-phase value (3.95%). Roostita and Fleet determined that the properties of yeasts that affected the spoilage rate of Camembert and blue-veined cheeses were the abilities to ferment/assimilate lactose, produce extracellular lipolytic and proteolytic enzymes, utilize lactic and citric acid, and grow at 10°C.

MICRO-ORGANISMS ACTION

The micro-organisms act in the milk and produce or change oudors, coloures flavours, gas enzymes and alkali. There are briefly explained here.

Odours

An acid odour can often be detected in milk having slight developed acidity. Micro-organisms acting upon both the proteins and the fat oflen produce changes which are detected by odour. Milk is frequently graded on the basis of its odours.

Colour

Moulds often discolour the surface of the butter and cheese, and may carry their discolouraion into the interior of these products. In the absence of proper sanitation, Pseudomonas spp may give rise to a yellowish film on product surfaces contact of equipment.

Flavour

The acid flavour commonly found in milk is caused by micro-organisms which attack the lactose. A bitter flavour results from certain types of protein breakdown caused by micro-organisms.

Mould contamination in milk products, mainly cream, butter, cheece and khoa may give undesired mouldy or nasty flavours. The quality of flavour in dahi, olher fermented milk, butter and cheesc is improved by the use of sele-cted micro-orgamisms.

Gas

Many micro-organisms, usually reaching with the lactose, produce gas in milk and milk products. Yeasts are particularly prone to gas, mainly carbon di-oxide. Gas formation is a common defect of sour cream. Gas production is essential in certain types of cheese such as, Emmanthal, and in one or two fermented milks such as kumiss.

Prevention of Spoilage in Milk

Enzymes

Some enzymes are produced in milk and milk products, but few of the natural enzymes of milk are attacked or changed by the action of micro-organisms.

Alkali

A few micro-organisms such as Alcaligenes viscolactis, Pseudomonas fluorescens, and Microccus ureae, produce alkali or neuteralise acidity developed in milk. Yeasts and moulds are seldom responsible for this action in milk.

Micro-organisms in Preparation of Milk Products

The following are the major milk product in the preparation of which micro- organism are involved.

Fermented Milk

Curd or dahi is a one of the fermented milk product. The species of lactic acid bacteria occuring most commonly in dahi include Lactobacillus bulgaricus, Streptococcus thermophilis, S. faecalis, S. lactis, L. casei and L. plantarum. The term "yoghurt" is widely used in other milk producing countries for products similar to duhi.

Streptococcus themlophillis and Lactobacililis bulgaricus are used together for making yoghurt, the former produces some acid and a fine aroma, the latter one produces high acidity. The wide variety of fermented milks includes not only those coagulated by acid formation, but also are characterised by flavours produced by bacteria and yeasts.

Cheese

The microtlora of cheese may differ greatly from one kind of cheese to another. The starter or inoculum used must vary accordingly. Here some of the varieties of bacteria and moulds employed in cheese making are mentioned. In cheddar cheese, the action of rennet is influenced by acidity. The starter used for this cheese, therefore, must contain acid producing bacteria. The most commonly used in this starter arc Streptococcus lactis or S. cremoris.

Roquefort or Blue cheese is made with the help of lactic acid starter and rennet, the bacteria is usually S. lactis. As the curd is dipped, spores of Penicillium roquefort (blue green mould) spreads throughout the curd. After being pressed the cheese is punctured to admit air to facilitate mould growth in its interior. Camembert cheese is similar to Roquefort. It is prepared in smaller prints and cured with P. camemberti which is a white mould.

Ghee

Ghee or clarified butter is extensively used in Indian homes. The micro biological problems encountered in its preservation are very few as it contains only a very little amount of moisture (less than 1%). It is almost pure butter fat and so is resistant against the attack of bacteria, except a few species of lipolytic organisms.

If the Ghee is not being packed under hygienic environment, it shows that it contains a large amount of bacteria. It contains micrococci, gram negative rods, spore forming mesophilic rods and moulds of the types of Aspergillus and Penicillium spp. The keeping quality of ghee depends primarily upon its moisture content.

Concentrated Milk Products

The two most common concentrated milk products found in many parts of India are khoa and rabri.

- *Khoa:* The keeping quality of khoa depends entirely upon the degree of contamination and the extent to which it is exposed to conditions favourable to the growth of micro-organisms subsequent to its manufacture. During the process of manufacture of khoa continued boiling process is followed. Almost all the bacteria present in milk arc killed, still a few spore formers may survive. Therefore, the bacteria present in khoa are mostly due to the faulty methods of its handling after preparation.
- *Rabri:* Rabri is another product of milk that is prepared by heating in a wide top open pan. In this case the milk is not boiled, but is heated to a temperature at which a skin may form on the surface. The skin which forms during heating is carefully removed from time to time by the use of a special dipper, usually made of bamboo. During the process of making, equipments used arc not sterilized and the bamboo dipper may act as a source of contamination. Besides these sources handling it after preparation also may result in contamination of the product. The most common organisms found in rabri are spore forming rods, micrococci, moulds and yeasts. Some disease producing bacteria may also enter into it.

Chhenna

Chhenna is one of the indigenous milk product in India; It is prepared from milk by precipitation of the protein, by the addition of an acid, either sour whey, citrus fruit juice or citric acid solution. The chhenna is then removed by draining off the whey. Chhenna also has a very low keeping quality and it tends to become sour rather rapidly.

It is highly perishable and its keeping quality is only 6 -8 hours in summer and 16-20 hours in winter. Though it is prepared after the milk has been boiled, spore forming aerobic rods survive boiling and they get the most favourable conditions for growth and at summer temperature they start growing very rapidly.

The types of bacteria present in chhenna represent micrococci, aerobic spore formers and non-spore forming rods. Mould growth on the surface is visible within 48 hours. The common moulds found in chhenna belong to the species of Penicillium aspergillus, Mucos, Rhizopus, Fusarium and Paecelomyces. Most of the bacterial species are thermoduric and the spore formers are all proteolytic. Moulds are also very active proteolytic agents.

Hygienically prepared chhenna may keep well for a longer period. It may be preserved by adding 0.5% sodium benzoate, or 0.15 to 0.2% sodium propionate.

In both the cases the keeping quality is increased up to 5 to 6 days. Sugar may also be added as a preservative. Thirty percent of sugar has been found to be more effective and it keeps well for 4 to 5 days. The wrapping materials also have great influence on the keeping quality of chhenna.

Mostly butter paper should be used for wrapping. It has also been noted that if the paper is rubbed with 0.5% solution of sodium benzoate or propionate the keeping quality is increased to a great extent.

Ice-Cream

Bacterial content of ice-cream is dependent on the quality of, the ingredients used. The ingredients used in ice-cream may include cream with or without added milk solids, sugar, gelatin, stabilizer, and eggs or egg solids.

Once the ingredients are frozen micro-organisms cannot multiply in it, Number of bacteria: Cream content in ice-cream varies widely in bacteriological condition. Pasteurization of the cream normally destroys a high percentage of the bacteria. Thermoduric bacteria, such as Micrococci and Stereptococci may be present in the cream. Utensils used to hold milk ice-cream mix should be thoroughly sterilized before using.

Egg products and gelatin may contribute large number of bacteria, including types of ice-creams. Sugar, in general, contains only small number of bacteria and is of little importance as a source of organisms in ice-cream.

Flavouring and colouring materials, fruits and nuts may constitute important sources of bacteria since, they often are added in the raw state just as the mix enters the freezer. These nuts carry a wide variety of organisms including aerobic Sporformers, Micrococci and E.coli.

Types of bacteria: A wide variety of bacterial species have been isolated from ice-cream.

These include Micrococci, Streptococci, aerobic and anaerobic spore formers, E. coli, A. aerogenes, Pseudomonas spp. and many other pathogenic and non-pathogenic bacteria. Acid forming species make up a considerable percentage of the flora. If favourable conditions for growth of bacteria are provided a variety of organisms grow in ice-creams and these include yeasts and moulds.

Disease transmission: Commercial ice-cream is known to cause many outbreaks of epidemics; Home-made ice-cream prepared with raw ingredients also may cause outbreaks of certain epidemics. Most of the outbreaks are due to food poisoning. Food poisoning is usually due to the growth of micrococcus Pyogenes aureus. It produces toxins 'that cause nausea, dizziness, headaches and diarrhoea.

Pasteurization of ice-cream mix destroys the Micrococci, but the toxins are not destroyed. Care should therefore, be taken to avoid contamination of the mix at high temperature, infection from the ice-cream makers and contamination from the utensils are some of the factors that are responsible for the outbreaks of diseases through ice-cream. A significant fact to be remembered is that if any disease-producing bacteria gets entry into the ice-cream it may be kept viable for years together in the frozen state.

Milk powder

Dried milk products represent the most stable products from the stand point of microbial spoilage. As long as they are kept in a dry state no micro-organisms of any type can develop in them. In fact the micro-

Prevention of Spoilage in Milk

organisms gradually decrease in numbers in dried foods during their storage. At present, milk is dried by the roller process and also by the various spray systems.

Since, the low content of moisture (less than 5%) absolutely prevents bacterial growth. In any storage temperature, the bacterial content of dried milk reflect the original position of milk and the conditions under which they were manufactured. Though the micro-organisms present In dried milk powder cannot multiply, they cause serious defects in the manufacture of different products. Bacillus mesentericus if present in the milk powder may cause a very serious defect known as ropy-bread originating from the use of such milk powder. When starter cultures arc prepared from dried milk, spore 10rmers 'present in it may give rise to serious troubles. Micro-organisms found in spray dried milk include thermoduric Streptococci such as S. themlophilus, micrococci, Lactobacilli, Achromobacter, aerobic and anaerobic spore forming types, moulds, yeast, coliform bacteria, and species of tiny rods known as microbacteria.

Butter

The quality of butter and its period of preservation depends greatly on the quality of the raw materials used in its preparation. The cream should be prepared from fresh clean milk produced and handled under hygienic conditions. Cream used for butter making is usually pasteurized at a temperature not below 85°C and therefore almost all the bacteria are destroyed except the spore formers.

Micro-organisms may gain entrance to cream after it has been pasteurized through the utensils and these organisms may include lactic acid bacteria, spore formers bacteria may gain entrance into butter through salt. Butter colour may act as a source of fungal mycelium in butter. However, butter colour and salt are not very great sources of bacteria in butter. Microflora of butter starter must contain S.lactis, S. cremoris, and the aroma producing diacetyl bacteria. In a starter the presence of S. lactis, S. cremoris, and S. diacetilactis is rnost favourable.

Colour Defects

Most of the colour defects in butter are caused by the microbial actions. Moulds that enter the butter from equipments, wrappers, or printers or from the air of the creamery or storage room. They produce brown, greenish, and black growth that discolours the surface of butter in addition to affecting the flavour.

Pseudomonas nigrifaciens enter butter from water and contaminate equipment. They produce a black pigment on the surface which may sometimes by mistaken for a grease smudge. Pink yeasts occasionally develop on the surface of butter and produce pink spots where they grow. Sulphur dioxide which sometimes escapes from the refrigerator may also cause pink discolouration on the surface of butter.

Moulds are generally present in the air and therefore it is very easy that butter gets contaminated by moulds. Cladosporium is generally found in abundance inside the butter and it is seen in dark patches. Mould growth in butter is an indication of the high moisture content in it. The higher the acidity of the butter more favourable is the conditions for the growth of moulds.

7

Food Applications for Human Health

QUANTITY AND QUALITY OF FOOD OILS

Food oils have both nutritional and functional qualities. From a nutritional perspective, fats and oils contribute more energy than any other nutrient category, about nine calories per gram. This compares with about four calories per gram from carbohydrates and protein. At the same time, specific fatty acids that comprise most of what we call "fat" can affect a person's risk of developing certain chronic diseases such as heart disease. Research over the past several decades has shown that some categories of fatty acids, such as saturated fatty acids, increase the risk of heart disease and other chronic diseases when consumed in excess. Fatty acids also influence how foods behave during manufacturing and processing. For example, saturated fatty acids add stability, texture, and flavour to foods, so they are not simple to replace.

To reduce the saturated fatty acid content of foods, plant breeders and food manufacturers increased their use of vegetable oils rich in polyunsaturated fatty acids and developed food oils low in saturated fatty acids. One example is canola oil with 6 per cent to 7 per cent total saturated fatty acids. To improve the stability of vegetable oils rich in polyunsaturated fatty acids, food manufacturers developed partially hydrogenated oils. The process of hydrogenation reduced the polyunsaturated fatty acid content and increased oil stability, but created *trans* fatty acids, which were subsequently associated with adverse health effects. As a result, hydrogenated fats, the main source of dietary *trans* fatty acids, are now being eliminated from foods. Food manufacturers are developing other ways to reduce undesirable saturated fat content while maintaining stability such as using short chain saturated fatty acids and monounsaturated fatty acids. To date, one functional food oil created with the tools of biotechnology has been commercialized. Calgene's high lauric acid canola, Laurical™, containing 38 per cent lauric acid, is used in confectionary products, chocolate, and non-food items such as shampoo. Conventional canola oil does not contain lauric acid. Laurical™ is a substitute for coconut and palm oils. FDA approved its use in foods in 1995. The following part describes research to date focused on developing crop varieties with other unique oil profiles.

Strategic Aims of Altered Fatty Acid Profile

Improving the healthfulness and functionality of food oils can be accomplished in several ways.

Where traditional plant breeding reaches its limits, biotechnology may be used to:

- Reduce saturated fatty acid content for "heart-healthy" oils
- Increase saturated fatty acids for greater stability in processing and frying
- Increase oleic acid in food oils for food manufacturing
- Reduce alpha-linolenic acid for improved stability in food processing
- Introduce various omega-3 polyunsaturated fatty acids including long-chain forms
- Enhance the availability of novel fatty acids, *e.g.,* gamma-linoleic acid

Quantity and Quality of Plant Protein

Efforts to improve the protein content and quality of staple foods have been underway for decades. The main focus is crops grown in developing countries, where nutrient shortfalls are widespread and dietary diversity limited. Foods such as potato and cassava, staple foods in several parts of South America and Africa, have less than one per cent protein. Efforts to improve protein quality strive to increase the amount of limiting essential amino acids provided by the protein in the food. The amino acids most often present in inadequate amounts are lysine, tryptophan, and methionine. Improvements in protein quality benefit both human and animal nutrition and increase the feed efficiency of crops fed to food animals. For example, corn is widely fed to cattle but it is limiting in lysine and methionine. Corn with higher levels of these amino acids would significantly improve feed efficiency and lower input costs to farmers.

Food Applications for Human Health

Improved corn varieties consumed by humans would also have nutritional benefits.

There are various ways of improving protein quantity and quality. One is to increase the total amount of protein produced by selecting germplasm with an altered balance of seed proteins. This may be done by traditional cross breeding or genetic engineering. Another approach is to introduce genes from other sources for proteins that have a favourable balance of essential amino acids. An example is the introduction into potato of a gene for seed albumin protein from amaranth. A third approach seeks to increase the production of specific amino acids such as lysine.

This approach was used in the development of Quality Protein Maize. William Folk and his team at the University of Missouri, Columbia, MO, pioneered another approach to improve seed protein quality. Their strategy was to substitute more desirable and scarce amino acids for more abundant ones in certain seed proteins. They applied this concept to rice by increasing the production of lysine, an essential amino acid, at the expense of the non-essential amino acids, glutamine, asparagine and glutamic acid. *Cassava*: A staple food for some 500 million people in tropical and sub-tropical parts of the world, cassava, also known as yucca or manioc, thrives in marginal lands having little rain and nutrient-poor soils. It is widely consumed in Africa, and parts of Asia and South America. Cassava root has less than 1 per cent protein and poor nutritional value.

However, the leaves are also consumed and these are a good source of beta-carotene, the precursor of vitamin A. In 2003, Zhang and colleagues reported using a synthetic gene to increase the protein content in cassava. The gene is for a storage protein rich in nutritionally essential amino acids. When the gene was expressed in cassava, transformed plants expressed the gene in roots and leaves, both of which are consumed in human diets.

The experiment demonstrated the feasibility of increasing the quantity and quality of protein in cassava. Cassava also contains cyanogenic glucosides that can produce chronic toxicity if not eliminated or reduced by grating, sun-drying, or fermenting. Efforts to develop cassava varieties low in these toxicants is a high research priority. *Corn*: Corn is the predominant staple food in much of Latin America and Africa. Although some varieties may contain appreciable quantities of protein, its quality is poor because of low lysine and tryptophan content. In 1964, it was discovered that corn bearing a gene known as opaque-2 contained increased concentrations of lysine and tryptophan and had significantly improved nutritional quality. However, opaque-2 corn proved to have low yields, increased susceptibility to diseases and pests, and inferior functional characteristics. At the International Maize and Wheat Improvement Centre in Mexico, work with the opaque-2 gene continued using both traditional breeding and molecular methods.

After at least 12 years' work, CIMMYT researchers succeeded in developing hardy corn varieties that contained twice the lysine and tryptophan content as traditional varieties, but were disease-resistant and high-yielding. Scientists Surinder K. Vasal and Evangelina Villegas of CIMMYT were awarded the World Food Prize in 2000 for their work developing 'Quality Protein Maize'. Quality Protein Maize varieties have been adapted to and released in over 40 countries in Latin America, Africa, and Asia. Recent researchers at CIMMYT reported the development of transgenic corn with multiple copies of the gene from amaranth that encodes for the seed storage protein amarantin. Total protein in the transgenic corn was increased by 32 per cent and some essential amino acids were elevated 8 per cent to 44 per cent. In 2004, a team of researchers at the University of California, Riverside, reported that transgenic corn with increased production of the plant regulating hormone, cytokinin, had nearly twice the content of protein and oil as conventional corn.

This development resulted from an unusual change in the way the plant developed. Normally, corn ears develop flowers in pairs, one of which usually dies. Under the influence of the additional cytokinin, both flowers developed but yielded only a single kernel. These kernels contained more protein and oil than conventional corn. Pursuing a different strategy to improve protein quality, researchers at Monsanto, St. Louis, MO, used genetic engineering techniques to reduce the amount of zein storage proteins. These storage proteins constitute over half the protein in corn and are deficient in lysine and tryptophan. Increased production of other proteins in the corn led to higher levels of lysine, tryptophan, and methionine. The agronomic and nutritional properties of these lines are currently being evaluated. Researchers at the Max Planck Institute, Germany, have focused on methionine, another limiting amino acid. They elucidated several key steps in methionine metabolism in plants.

This work, currently in the preliminary stage, could pave the way for using genetic engineering techniques to improve the methionine content of plants. *Potato*: Potato is a dietary staple throughout parts of Asia, Africa, and South America. Typically, potatoes contain about 2 per cent protein and 0.1 per cent fat. It was reported in 2000 that, as in cassava, transfer of the gene for seed albumin protein from *Amaranthus hypochondriacus* to potato resulted in a "striking" increase in protein content of the transgenic potatoes. In 2004, researchers at the National Centre for Plant Genome Research, India, reported the development of a nutritionally improved potato line with 25 per cent higher yields of tubers and 35 per cent–45 per cent greater protein content. Dubbed the "protato," the protein-rich potato had significant increases in lysine and methionine, which enhance the quality of the additional protein. In February 2004, this potato was reported "approaching release" to farming communities. It should be noted that while potatoes are known for their high starch content, it has been possible to genetically engineer potatoes that contain fat. In July 2004, Klaus and colleagues at the Max Planck Institute of Molecular Plant Physiology demonstrated increased fatty acid synthesis in potatoes. *Rice*: Almost half the world's population eats rice, at least once a day. Rice is the staple food among the world's poor, especially in Asia and parts of Africa and South America. It is the primary source of energy and nutrition for millions.

Thus, improving the nutritional quality of rice could potentially improve the nutritional status of nearly half the world's population, particularly its children. Commodity rice contains about 7 per cent protein, but some varieties, notably black rice, contain as much as 8.5 per cent. The most limiting amino acid in rice is lysine. Efforts to increase the nutritional value of rice target protein content and quality along with key nutrients often deficient in rice-eating populations, such as vitamin A and iron.

The International Rice Research Institute, Philippines, is a primary centre for rice research and development of improved varieties. In 1999, Dr. Momma and colleagues at Kyoto University, Japan, reported a genetically engineered rice having about 20 per cent greater protein content compared with control rice. Transgenic plants containing a soybean gene for the protein glycinin contained 8.0 per cent protein and an improved essential amino acid profile compared with 6.5 per cent protein in the control rice. Dr. William Folk and his team genetically modified rice to increase its content of the amino acid lysine. They did so by modifying the process of protein synthesis, rather than by gene transfer or the expression of new proteins. They achieved an overall 6 per cent increase in lysine content in the grain. Although lysine content remained below optimum levels, the scientists suggested that additional transformations and modifications could further boost lysine levels.

Perhaps the most famous genetic transformations in rice are those in "Golden Rice" involving the vitamin A precursor, beta-carotene, and iron.

The lead scientist in the golden rice project, Dr. Ingo Potrykus, now retired from the Swiss Federal Institute of Technology, was also involved in applying biotechnology for the improvement of rice protein. Although details are sparse, Potrykus described the work of Dr. Jesse Jaynes, who synthesized a synthetic gene coding for an ideal high-quality storage protein with a balanced mixture of amino acids. The gene, named Asp-1, was transferred to rice with the appropriate genetic instructions for its production in the endosperm or starchy part of the rice grain.

The transgenic rice plants accumulated the Asp-1 protein in their endosperm in a range of concentrations and provided essential amino acids but data are not yet available on the concentrations achieved or their nutritional relevance. Precedent for the expression of a synthetic gene in rice grown in cell culture suggests that Jaynes' approach is viable.

BIOTECHNOLOGY AND GLOBAL HEALTH

The World Health Organization estimates that more than 8 million lives could be saved by 2010 by combating infectious diseases and malnutrition through developments in biotechnology. A study conducted by the Joint Centre for Bioethics at the University of Toronto identified biotechnologies with the greatest potential to improve global health, including the following:

- Hand-held devices to test for infectious diseases including HIV and malaria. Researchers in Latin America have already made breakthroughs with such devices in combating dengue fever.
- Genetically engineered vaccines that are cheaper, safer, and more effective in fighting HIV/AIDS, malaria, tuberculosis, cholera, hepatitis, and other ailments. Edible vaccines could be incorporated into potatoes and other foods.
- Drug delivery alternatives to needle injections, such as inhalable or powdered drugs.
- Genetically modified bacteria and plants to clean up contaminated air, water, and soil.
- Vaccines and microbicides to help prevent sexually transmitted diseases in women.
- Computerized tools to mine genetic data for indications of how to prevent and cure diseases.

- Genetically modified foods with greater nutritional value.

Labelling of genetically modified foods has sparked additional debate. Labels are required on food produced through biotechnology to inform consumers of any potential health or safety risk. For example, a label is required if a potential allergen is introduced into a food product. A label is also required if a food is transformed so that its nutrient content no longer resembles the original food. For example, so-called golden rice has been genetically engineered to have a higher concentration of beta-carotene than regular rice, and thus it must be included on the label. In response to consumer demands, regulators in England have instituted mandatory labelling laws for all packaged foods and menus containing genetically modified ingredients. Similar but less restrictive laws have been instituted in Japan. In Canada, the policy on labelling has remained similar to that of the United States.

Some consumer advocates maintain that not requiring a label on all genetically modified foods violates consumers' right to make informed food choices, and many producers of certain foods, such as foods containing soy protein, now include the term "non-GMO" on the label to indicate that the product does not contain genetically modified ingredients.

The application of recombinant DNA technology to foods, commonly called biotechnology, may be viewed as an extension of traditional crossbreeding and fermentation techniques. The technology enables scientists to transfer genetic material from one species to another, and may produce food crops and animals that are different than those obtained using traditional techniques. The FDA has established procedures for approval of food products manufactured using recombinant DNA technology that require food producers to demonstrate the safety of their products. The American Dietetic Association, the American Medical Association, and the World Health Organization have each adopted statements that techniques of biotechnology may have the potential to improve the food supply. These organizations and others acknowledge that long-term health and environmental impacts of the technology are not known, and they encourage continual monitoring of potential impacts.

Genetically Modified Organism

A genetically modified organism (GMO) or genetically engineered organism (GEO) is an organism whose genetic material has been altered using genetic engineering techniques. These techniques, generally known as recombinant DNA technology, use DNA molecules from different sources, which are combined into one molecule to create a new set of genes. This DNA is then transferred into an organism, giving it modified or novel genes. Transgenic organisms, a subset of GMOs, are organisms which have inserted DNA that originated in a different species. Some GMOs contain no DNA from other species and are therefore not transgenic but cisgenic.

History

The general principle of producing a GMO is to add new genetic material into an organism's genome. This is called genetic engineering and was made possible through the discovery of DNA and the creation of the first recombinant bacteria in 1973, *i.e.*, *E.coli* expressing a Salmonella gene.

This led to concerns in the scientific community about potential risks from genetic engineering, which were thoroughly discussed at the Asilomar Conference. One of the main recommendations from this meeting was that government oversight of recombinant DNA research should be established until the technology was deemed safe. Herbert Boyer then founded the first company to use recombinant DNA technology, Genentech, and in 1978 the company announced creation of an *E. coli* strain producing the human protein insulin. In 1986, field tests of bacteria genetically engineered to protect plants from frost damage (ice-minus bacteria) at a small biotechnology company called Advanced Genetic Sciences of Oakland, California, were repeatedly delayed by opponents of biotechnology. In the same year, a proposed field test of a microbe genetically engineered for a pest resistance protein by Monsanto Company was dropped.

Production

Genetic modification involves the insertion or deletion of genes. When genes are inserted, they usually come from a different species, which is a form of horizontal gene transfer. In nature this can occur when exogenous DNA penetrates the cell membrane for any reason. To do this artificially may require attaching the genes to a virus or just physically inserting the extra DNA into the nucleus of the intended host with a very small syringe, or with very small particles fired from a gene gun. However, other methods exploit natural forms of gene transfer, such as the ability of *Agrobacterium* to transfer genetic material to plants, or the ability of lentiviruses to transfer genes to animal cells.

Uses

GMOs have widespread applications. They are used in biological and medical research, production of pharmaceutical drugs, experimental medicine (*e.g.* gene therapy), and agriculture (*e.g.* golden rice). The term "genetically modified organism" does not always imply, but can include, targeted insertions of genes from one species into another. For example, a gene from a jellyfish, encoding a fluorescent protein called GFP, can be physically linked and thus co-expressed with mammalian genes to identify the location of the protein encoded by the GFP-tagged gene in the mammalian cell. Such methods are useful tools for biologists in many areas of research, including those who study the mechanisms of human and other diseases or fundamental biological processes in eukaryotic or prokaryotic cells.

To date the broadest application of GMO technology is patent-protected food crops which are resistant to commercial herbicides or are able to produce pesticidal proteins from within the plant, or *stacked trait* seeds, which do both. The largest share of the GMO crops planted globally are owned by Monsanto Company, according to the company. In 2007, Monsanto's trait technologies were planted on 246 million acres (1,000,000 km^2) throughout the world, a growth of 13 percent from 2006. In the corn market, Monsanto's triple-stack corn – which combines Roundup Ready 2 weed control technology with YieldGard Corn Borer and YieldGard Rootworm insect control – is the market leader in the United States. U.S. corn farmers planted more than 17 million acres (69,000 km^2) of triple-stack corn in 2007, and it is estimated the product could be planted on 45 million to 50 million acres (200,000 km^2) by 2010. In the cotton market, Bollgard II with Roundup Ready Flex was planted on nearly 3 million acres. Rapid growth in the total area planted is measurable by Monsanto's growing share. On January 3, 2008, Monsanto Company (MON.N) said its quarterly profit nearly tripled, helped by strength in its corn seed and herbicide businesses, and raised its 2008 forecast.

According to the International Service for the Acquisition of Agri-Biotech Applications (ISAAA), of the approximately 8.5 million farmers who grew biotech crops in 2005, some 90 per cent were resource-poor farmers in developing countries. These include some 6.4 million farmers in the cotton-growing areas of China, an estimated 1 million small farmers in India, subsistence farmers in the Makhathini flats in KwaZulu Natal province in South Africa, more than 50,000 in the Philippines and in seven other developing countries where biotech crops were planted in 2005.. ISAAA estimated that by 2008, 13.3 million farmers were growing GM crops, including 12.3 million in developing counties. These comprised 7.1 million in China (Bt cotton), 5.0 million in India (Bt cotton), and 200,000 in the Philippines.

"The Global Diffusion of Plant Biotechnology: International Adoption and Research in 2004", a study by Dr. Ford Runge of the University of Minnesota, estimates the global commercial value of biotech crops grown in the 2003–2004 crop year at US$44 billion.

In the United States the United States Department of Agriculture (USDA) reports on the total area of GMO varieties planted. According to National Agricultural Statistics Service, the States published in these tables represent 81-86 percent of all corn planted area, 88-90 percent of all soybean planted area, and 81-93 percent of all upland cotton planted area (depending on the year). USDA does not collect data for global area. Estimates are produced by the International Service for the Acquisition of Agri-biotech Applications (ISAAA) and can be found in the report, Global Status of Commercialized Transgenic Crops: 2007. Transgenic animals are also becoming useful commercially. On 6 February 2009 the U.S. Food and Drug Administration approved the first human biological drug produced from such an animal, a goat. The drug, ATryn, is an anticoagulant which reduces the probability of blood clots during surgery or childbirth. It is extracted from the goat's milk.

Detection

Testing on GMOs in food and feed is routinely done by molecular techniques like DNA microarrays or qPCR. The test can be based on screening elements or event-specific markers for the official GMOs (like Mon810, Bt11, or GT73). The array-based method combines multiplex PCR and array technology to screen samples for different potential GMOs, combining different approaches (screening elements, plant-specific markers, and event-specific markers). The qPCR is used to detect specific GMO events by usage of specific primers for screening elements or event-specific markers.

To avoid any kind of false positive or false negative testing outcome, comprehensive controls for every step of the process is mandatory. A CaMV check is important to avoid false positive outcomes based on virus contamination of the sample.

Food Applications for Human Health

Transgenic Animals

Transgenic animals are used as experimental models to perform phenotypic tests with genes whose function is unknown. Genetic modification can also produce animals that are susceptible to certain compounds or stresses for testing in biomedical research. Other applications include the production of human hormones such as insulin. In biological research, transgenic fruit flies (*Drosophila melanogaster*) are model organisms used to study the effects of genetic changes on development.

Fruit flies are often preferred over other animals due to their short life cycle, low maintenance requirements, and relatively simple genome compared to many vertebrates. Transgenic mice are often used to study cellular and tissue-specific responses to disease.

This is possible since mice can be created with the same mutations that occur in human genetic disorders, the production of the human disease in these mice then allows treatments to be tested. In 2009 scientists in Japan announced that they had successfully transferred a gene into a primate species (marmosets) and produced a stable line of breeding transgenic primates for the first time.

It is hoped that this will aid research into human diseases that cannot be studied in mice, for example Huntington's disease and strokes.

Cnidarians such as *Hydra* have become attractive model organisms to study the evolution of immunity. For analytical purposes an important technical breakthrough was the development of a transgenic procedure for generation of stably transgenic hydras by embryo microinjection.

Transgenesis in fish with promoters driving an overproduction of "all fish" growth hormone has resulted in dramatic growth enhancement in several species, including salmonids, carps and tilapias. These fish have been created for use in the aquaculture industry to increase the speed of development and potentially, reduce fishing pressure on wild stocks. None of these GM fish have yet appeared on the market, mainly due to the concern expressed among the public of the fish's potential negative effect on the ecosystem should they escape from fish farms.

Transgenic Microbes

Bacteria were the first organisms to be modified in the laboratory, due to their simple genetics. These organisms are now used for several purposes, and are particularly important in producing large amounts of pure human proteins for use in medicine. Genetically modified bacteria are used to produce the protein insulin to treat diabetes. Similar bacteria have been used to produce clotting factors to treat haemophilia, and human growth hormone to treat various forms of dwarfism. These recombinant proteins are safer than the products they replaced, since the older products were purified from cadavers and could transmit diseases. Indeed the human-derived proteins caused many cases of AIDS and hepatitis C in haemophilliacs and Creutzfeldt-Jakob disease from human growth hormone.

For instance, the bacteria which cause tooth decay are called *Streptococcus mutans*. These bacteria consume leftover sugars in the mouth, producing lactic acid that corrodes tooth enamel and ultimately causes cavities. Scientists have recently modified *Streptococcus mutans* to produce no lactic acid. These transgenic bacteria, if properly colonized in a person's mouth, could reduce the formation of cavities. Transgenic microbes have also been used in recent research to kill or hinder tumours, and to fight Crohn's disease. Genetically modified bacteria are also used in some soils to facilitate crop growth, and can also produce chemicals which are toxic to crop pests.

Gene Therapy

Gene therapy, uses genetically modified viruses to deliver genes that can cure disease into human cells. Although gene therapy is still relatively new, it has had some successes. It has been used to treat genetic disorders such as severe combined immunodeficiency, and treatments are being developed for a range of other currently incurable diseases, such as cystic fibrosis, sickle cell anemia, and muscular dystrophy.. Current gene therapy technology only targets the non-reproductive cells meaning that any changes introduced by the treatment can not be transmitted to the next generation. Gene therapy targeting the reproductive cells-so called "Germ line Gene Therapy"-is very controversial and is unlikely to be developed in the near future.

Transgenic Plants

Transgenic plants have been engineered to possess several desirable traits, including resistance to pests, herbicides or harsh environmental conditions, improved product shelflife, and increased nutritional value. Since the first commercial cultivation of genetically modified plants in 1996, they have been modified to be tolerant to the herbicides glufosinate and glyphosate, to be resistant to virus damage as in Ringspot virus resistant GM papaya, grown in Hawaii, and to produce the Bt toxin, a potent insecticide.

Bt-maize is a corn that has been genetically modified by splicing the toxin-producing gene from bacteria into the DNA sequence of the corn in order to sicken or kill insects that try to consume it. While some genetically modified crops are more nutritious because they contain these extra vitamins and minerals, they will not cure all of the malnutrition-related ailments in the world and should only be a supplement to a balanced diet.

Cisgenic Plants

Genetically modified sweet potatoes have been enhanced with protein and other nutrients, while golden rice, developed by the International Rice Research Institute, has been discussed as a possible cure for Vitamin A deficiency. In reality, customers would have to eat twelve bowls of rice a day in order to meet the recommended levels of Vitamin A. In January 2008, scientists altered a carrot so that it would produce calcium and become a possible cure for osteoporosis; however, people would need to eat 1.5 kilograms of carrots per day to reach the required amount of calcium.

The coexistence of GM plants with conventional and organic crops has raised significant concern in many European countries. Since there is separate legislation for GM crops and a high demand from consumers for the freedom of choice between GM and non-GM foods, measures are required to separate foods and feed produced from GMO plants from conventional and organic foods. European research programmes such as Co-Extra, Transcontainer and SIGMEA are investigating appropriate tools and rules. At the field level, biological containment methods include isolation distances and pollen barriers.

Biological Process

The use of GMOs has sparked significant controversy in many areas. Some groups or individuals see the generation and use of GMO as intolerable meddling with biological states or processes that have naturally evolved over long periods of time, while others are concerned about the limitations of modern science to fully comprehend all of the potential negative ramifications of genetic manipulation.

Foodchain

The safety of GMOs in the foodchain has been questioned, with concerns such as the possibilities that GMOs could introduce new allergens into foods, or contribute to the spread of antibiotic resistance. Although scientists have assured consumers of the safety of these types of crops, consumption has been discouraged in many countries by food and environmental activist groups who protest GM crops, claiming they are unnatural and therefore unsafe. This has led to the adoption of laws and regulations that require safety testing of any new organism produced for human consumption.

Trade with Europe and Africa

In response to negative public opinion, Monsanto announced its decision to remove their seed cereal business from Europe, and environmentalists crashed a World Trade Organization conference in Cancun that promoted GM foods and was sponsored by Committee for a Constructive Tomorrow (CFACT). Some African nations have refused emergency food aid from developed countries, fearing that the food is unsafe. During a conference in the Ethiopian capital of Addis Ababa, Kingsley Amoako, Executive Secretary of the United Nations Economic Commission for Africa (UNECA), encouraged African nations to accept genetically modified food and expressed dissatisfaction in the public's negative opinion of biotechnology.

RISK OF GMOS AND GM FOODS TO HUMAN HEALTH AND THE ENVIRONMENT

Introduction of a transgene into a recipient organism is not a precisely controlled process, and can result in a variety of outcomes with regard to integration, expression and stability of the transgene in the host.

History of Risk Assessment of GMOs

When new foods are developed by traditional breeding methods, they are usually not subject to specific pre- or postmarket risk or safety assessment by national authorities or through international standards. This is in contrast to requirements introduced for GMOs and GM foods. The concept of risk assessment of GMOs was first discussed at the Asilomar Conference in 1975. The discovery of recombinant DNA had raised concerns among researchers regarding the potential creation of recombinant viruses whose escape would threaten public health. Fourteen months after a voluntary moratorium on research involving recombinant DNA techniques, guidelines for the physical and biological containment of riskier experiments were drafted and agreed. These guiding principles were the basis of the USA guidelines for research in modern biotechnology developed in 1976 by the National Institutes of Health Recombinant DNA Advisory Committee. Other countries were soon to follow. Early regulatory requirements were intended to prevent the accidental release of microorganisms from research facilities. In continuation of this, regulation for contained use and deliberate release of GMOs was developed, *e.g.* EU regulations in 1990. These guidelines elaborated a premarket human-health and environmental-safety assessment requirement for all GMOs and GM foods on the basis that they are novel and have no history of safe food or environmental use. Many countries have since established specific premarket regulatory systems requiring the rigorous assessment of GMOs and GM foods before their release into the environment and/or use in the food supply.

A summary of some national and international legislation is available on the OECD Internet site. While many national regulatory bodies base their safety assessment of GMOs and GM foods on shared concepts, differences in regulatory systems have led to disagreements and confusion in their deployment. While the terms 'safety assessment' and 'risk assessment' are often used interchangeably in some literature, these are two clearly different, but interlinked processes. To provide international consistency in risk analysis of GMOs and GM foods which incorporates risk assessment, management and communication components, a number of international regulatory and standard-setting bodies have introduced uniform standards.

These include standards for human-health and environmental-safety assessment of GMOs and GM foods, and notification of their movement across national borders. The objective of uniform global standards for risk assessment would be challenging as countries are bound to reach different decisions on the scope of the assessment, particularly the resolution of whether or not to include social or economic aspects. International regulatory systems covering GM food safety and environmental safety came into force in 2003. The concept that allows for the comparison of a final product with one having an acceptable standard of safety is an important element of a GM food safety assessment. This principle was elaborated by FAO, WHO and OECD in the early 1990s and referred to as 'substantial equivalence'.

The principle suggests that GM foods can be considered as safe as conventional foods when key toxicological and nutritional components of the GM food are comparable to the conventional food and when the genetic modification itself is considered safe.

However, the concept has been criticized by some researchers. At a *Joint FAO/WHO consultation on foods derived from biotechnology* held in 2000, it was acknowledged that the concept of substantial equivalence contributes to a robust safety assessment, but it was also clarified that the concept should represent the starting point used to structure the safety assessment of a GM food relative to its conventional counterpart.

The consultation concluded that a consideration of compositional changes should not be the sole basis for determining safety, and that safety can only be determined when the results of all aspects under comparison are integrated.

This study does not cover aspects of occupational health which are often addressed in regulations dealing with the safety of work with GMOs in contained areas. It should also be noted that the adventitious presence of non-approved products of modern biotechnology among the approved is not within the scope of this study.

Assessment of the impact of GM foods on human health

Principles for the Safety Assessment of GM Foods

The Codex Alimentarius Commission adopted the following texts in July 2003: *Principles for the risk analysis of foods derived from modern biotechnology*; *Guideline for the conduct of food safety assessment of foods derived from recombinant-DNA plants*; and *Guideline for the conduct of food safety assessment of foods produced using recombinant-DNA microorganisms*. The last two texts are based on the *Principles* and describe methodologies for conducting safety assessments for foods derived from recombinant-DNA plants and microorganisms, respectively. The premise of the *Principles* dictates a premarket assessment, performed on a case-by-case basis and including an evaluation of both direct effects and unintended effects.

The Codex safety assessment principles for GM foods require investigation of:
- Direct health effects;
- Tendency to provoke allergic reactions;
- Specific components thought to have nutritional or toxic properties;
- Stability of the inserted gene;

Food Applications for Human Health

- Nutritional effects associated with the specific genetic modification; and
- Any unintended effects which could result from the gene insertion.

Codex principles do not have a binding effect on national legislation, but are referred to specifically in the *Agreement on the Application of Sanitary and Phytosanitary Measures* of the World Trade Organization and are often used as a reference in the case of trade disputes.

The 2003 *Expert consultation on the safety assessment of foods derived from GM animals, including fish* formed the opinion that to further develop the risk-assessment process with current scientific knowledge, integrated toxicological and nutritional evaluations should be conducted in order to identify food-safety issues that may need further investigation. Both evaluations combine data from the hazard identification and characterization, and food intake assessment steps.

It should be noted that such newly suggested further developments of the risk-assessment process have not yet been considered by Codex, and that the international principles and guidelines for risk analysis and safety assessment of foods derived from biotechnology are as accepted by Codex in 2003.

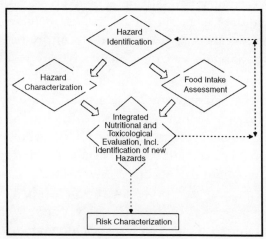

Fig. Schematic overview of a Suggested Further Development of the Risk Assessment Process

Potential Direct Effects on Human Health

The potential direct health effects of GM foods are generally comparable to the known risks associated with conventional foods, and include, for example, the potential for allergenicity and toxicity of components present, and the nutritional quality and microbiological safety of the food. Many of these issues have not traditionally been specifically assessed for conventional food; but in one area—toxicity of food components—there is ample experience related to the use of animal experiments to test potential toxicity of targeted chemical components. However, the intrinsic difficulty in testing whole foods, as opposed to specific components, in animal feeding experiments have resulted in the development of alternative approaches for the safety assessment of GM foods. The safety assessment of GM food follows a stepwise process aided by a series of structured questions.

Factors taken into account in the safety assessment include:
- Identity of gene of interest, including sequence analysis of flanking regions and copy number;
- Source of gene of interest;
- Composition of GMO;
- Protein expression product of the novel DNA;
- Potential toxicity;
- Potential allergenicity; and
- Possible secondary effects from gene expression or the disruption of the host DNA or metabolic pathways, including composition of critical macronutrients, micronutrients, antinutrients, endogenous toxicants, allergens and physiologically active substances.

A series of FAO/WHO expert consultations held in 2000, 2001 and 2003 recognized that animal studies can be of help but that there are practical difficulties in obtaining meaningful information from conventional toxicology testing, especially with whole-food studies in laboratory animals. The consultations also noted that very little is known about the potential long-term effects of any foods. At present, there is no conclusive information available on the possible health effects of modifications which would significantly change the nutritional characteristics of any food, such as nutritionally enhanced foods.

Potential Unintended Effects of GM Foods on Human Health

Unintended effects, such as elevated levels of antinutritional or toxic constituents in food, have on occasion been characterized in conventional breeding methods, *e.g.* glycoalkaloid levels in potatoes. Organisms derived from conventional breeding methods, including tissue cultures, may have a somewhat enhanced possibility for genetic instabilities, such as the activity of mobile elements and gene-silencing effects. These effects could increase the probability of unintended pleiotropic effects, *e.g.* increased or decreased expression of constituents or possibly modifications in expressed proteins, as well as epistasis. It has been argued that random insertion of genes in GMOs may cause genetic and phenotypic instabilities but, as yet, no clear scientific evidence for such effects is available. A better understanding of the impact of natural transposable elements on the eukaryotic genome may shed some light on the random insertion of sequences. Gene expression in conventional and GM crops is subject to environmental influences. Environmental conditions such as drought or heat can stimulate some genes; turning the expression up or down.

The assessment of potential synergistic effects is necessary in the risk assessment of organisms derived from gene stacking, *i.e.* breeding of GMOs containing genetic constructs with multiple traits. Internationally agreed procedures for the assessment of such organisms are desirable. Unintended effects can be classified as insertional effects, *i.e.* related to the position of insertion of the gene of interest, or as secondary effects, associated with the interaction between the expressed products of the introduced gene and endogenous proteins and metabolites.

There is common agreement that targeted approaches, *i.e.* the measuring of single compounds, is very useful and adequate to detect such effects, as has been done with conventionally bred products. To enhance and improve the identification and analyses of these unintended effects, profiling methods have been suggested. This untargeted approach allows detection of unintended effects at the mRNA, protein and metabolite level. It still remains to be seen which of these techniques would be useful for routine risk-assessment purposes. Unintended effects were specifically addressed by the FAO/WHO Expert consultation on the safety aspects of genetically modified foods of plant origin and the Codex Principles for the risk analysis of foods derived from modern biotechnology.

These consultations noted that there is a need to establish the consequences of natural baseline variations, the effects of growing conditions and environmental influences, and the ways to interpret safety-relevant data from profiling techniques. Adequate methods for the assessment of potential, unintended effects need to be evaluated for specific GMOs case by case, where the assessment already aims to consider unintended toxic and antinutritional factors through analysis of proximal constituents and GM characteristics. As profiling methods are not in use in routine risk assessment, the second step in the comparative safety assessment has been suggested as a measure for identifying and characterizing any unintended effects that may be associated with complex foods.

Potential Human-Health Effects from Horizontal Gene Transfer

Natural genetic transformation has been found to occur in different environments, *e.g.* in food. In addition, it has been shown that ingested DNA from food is not completely degraded by digestion, and that small fragments of DNA from GM foods can be found in different parts of the gastrointestinal tract. As the consequences of horizontal gene transfer may be significant in some human-health conditions, the potential for HGT needs to be part of the risk assessment of GM food. FAO/WHO consultations have also discussed the potential risks of gene transfer from GM foods to mammalian cells or gut bacteria.

These panels have suggested that it may be prudent in a food-safety assessment to assume that DNA fragments survive in the human gastrointestinal tract and can be absorbed by either the gut microflora or somatic cells lining the intestinal tract. It was agreed that the assessment needs to take into account a number of factors including, but not limited to, the specific characteristics encoded by the DNA sequences, the characteristics of the receiving organism, and the selective conditions of the local environment of the receiving organisms. Some scientists have pointed to the present methodological limitations of a comprehensive scientific evaluation of this problem. Discussion also addresses the consequences of a rare probability of a transfer event against the high numbers of bacteria and genes available for transfer.

The DNA construct used to change the genetic composition of a recipient organism should be considered within an assessment, especially if the gene or its promoter has been derived from a viral source. Sequences unrelated to the target gene could be introduced as part of the construct. Inadvertent introduction of such sequences into the germ-line of a GM animal not only has the potential for creating unintended genetic damage, but can also contribute by recombination to the generation of novel infectious viruses. A well-known example is the generation of a replication-competent murine leukaemia virus during the development of a vector containing a globin gene.

The horizontal transfer of recombinant genetic material to microorganisms has demonstrated an enhanced stability of DNA under certain conditions. Natural transformation of DNA to bacteria involves

Food Applications for Human Health

the active uptake of extracellular DNA by bacteria in a status of competence or in rare, illegitimate recombination events.

The probability of such an event occurring appears to be extremely low, and very much related to the genes, constructs and organisms in question.

The FAO/WHO expert panels concluded that horizontal gene transfer is a rare event that cannot be completely discounted, and that the consequences of such transfer should be considered in a safety assessment.

The panels encouraged the use of recombinant DNA without antibiotic-resistance genes or any other sequences which could stimulate transfer. The panels also discouraged the use of any unnecessary DNA sequences, including marker genes in the genetic construct. The safety assessment of a genetic construct should also examine the included marker genes.

Commonly used marker genes code for antibiotic resistance. Risk assessment of these selectable genes should focus on gene transfer to microorganisms residing in the gastrointestinal tract of humans or animals. As the potential of this gene transfer cannot be completely ruled out, the safety assessment should also consider information on the role of the antibiotic in human and veterinary medical uses.

Potential Immune Responses and Allergenicity Iinduced by GM Foods

Food allergies or hypersensitivities are adverse reactions to foods triggered by the immune system. Within the different types of reactions involved, non-immunological intolerances to food and reactions involving components of the immune system need to be differentiated. The former may invoke reactions such as bloating or other unpleasant reactions, but are thought not to involve the immune system and called 'food intolerances'.

Allergic reactions to traditional foods are well known. The major food allergens are proteins in and derived from eggs, fish, milk, peanuts, shellfish, including crustaceans and molluscs, soy, tree nuts and wheat. Whereas the groups of main allergens are well known and advanced testing methods have been elaborated, traditionally developed foods are not generally tested for allergens before market introduction. The application of modern biotechnology to crops has the potential to make food less safe if the newly added protein proves to cause an allergic reaction once in the food supply. A well-known case is the transfer of a gene encoding a known allergen, the 2S-Albumin gene from the Brazil nut, to a previously safe soybean variety.

When the allergenic properties of the transgenic soybean were tested, sera from patients allergic to Brazil nuts cross-reacted with the transgenic soybean. For this reason, a commercial product was never pursued. On the other hand, the introduction of an entirely new protein that has not been previously found in the food chain represents a different case. In the first case, guidelines for assessing foods with known allergens are clear. The second case is more difficult to assess because there is no definitive test to determine the potential allergenicity of a novel protein. Instead, several risk factors provide a rough guide as to the likelihood of allergenicity.

Risk-assessment protocols for food allergy examine four elements:
1. Allergenicity assessment
2. Dose response assessment
3. Exposure assessment and
4. Susceptible subpopulations.

Elements of an allergenicity assessment include a comparison of the sequence of the transferred gene with sequence motifs of allergenic proteins from databanks, an evaluation of the stability of the newly expressed proteins against digestion, and animal and immune tests, as appropriate. Absence of sequence similarity with allergenic protein epitopes, and low stability under acidic or proteolytic conditions, do not preclude the presence of a potential allergen. There are proven incidents which have contradicted the general rules, *e.g.* where small modifications in a protein sequence determine allergenicity. Allergenicity prediction using protein-sequence motifs identified from a new allergen database has been proposed as a new and superior strategy for identifying potential allergens. Some experts consider that the use of sera from polysensitized patients is important for the testing of allergenicity.

Areas of improvement of risk assessment of allergens include mechanistic studies of animal models and genomic techniques. FAO/WHO expert panels have established protocols for evaluating the allergenicity of GM foods on the basis of the weight of evidence. The strategy adopted is applicable to foods containing a gene derived from either a source known to be allergenic or a source not known to be allergenic. The panels have, however, discouraged the transfer of genes from known allergenic foods unless it can be demonstrated that the protein product of the transferred

Food Applications for Human Health

gene is not allergenic. These principles have been applied by many regulatory agencies assessing the safety of GM foods and have provided the basis for Codex guidelines for the safety assessment of foods derived from biotechnology. The cellular basis of immune responses is not completely understood, and a better understanding of the interaction of the immune system and foods in general is required in order to decipher whether specific GM foods may have impacts on the immune system apart from allergenicity. The impact of cell-mediated reactions on hypersensitivity reactions elicited by foods is a matter of current research.

Safety Aspects of Food Derived from GM Animals

Genetically modified animals have mainly been produced for biomedical research purposes. To date, no GM food animals have been introduced onto international markets. But GM food animals such as fish can be expected in the near future. In principle, the assessment of food and feed safety for GM animals follows the general principles of the assessment of GMOs. However, the specificities of the introduction of transgenes into animals, often using viral constructs for introduction into the germ-line, need distinct consideration. A 2003 report of the Pew Initiative on Food and Biotechnology reviewed techniques for the production, uses and welfare of GM animals, as well as safety aspects.

The risk assessment of foods derived from GM animals needs to be undertaken, as for other GM foods, on a case-by-case basis. This includes an assessment of potential recombination of viral vectors used for transformation with wild-type viruses, especially in poultry, where potential incomplete digestion could lead to intestinal uptake of orally administered proteins, and an assessment of peptide expression that may have hormonal activity. The FAO/WHO expert consultation on the *Safety assessment of foods derived from GM animals, including fish* held in 2003 addressed the key issues for food safety and evaluated the extent of scientific knowledge with regard to hazard identification and characterization unique to transgenic animals.

Phenotypic Analysis

Because of their size, and limitations in the generation process, there will in general be few initial founders for screening of GM animals, meaning that information on the variation range between animals with the same genetic modification will be rather limited. This will make interpretation of differences difficult. Furthermore, a selection of the edible tissues and products to be analysed has to be made for the different animal species. In specific cases, phenotypic analysis may also be advisable after processing or, for fish, during the various stages of spoilage. For example, adverse biogenic amines can be formed during spoilage in salmon, tuna, herring and other fish species. Similarly, formaldehyde may be produced in spoiled shrimp, cod, hake and many other species.

Compositional Analysis

Background data on the natural variation for individual constituents in different tissues need to be generated. Data in existing databases must be evaluated for their quality and value for use in comparative compositional analysis.

Safety Aspects of Foods Derived from or Produced with GMMs

The production of food additives or processing aids using GMMs, where the microorganism is not a part of the food, has become an important and generally well-accepted technology, with a significant number of such products on the market. Experience with the purification of proteins in the biomedical field suggests that well-standardized purification protocols are of central importance for the safety of these products. Where the GMMs are a part of the food matrix, certain criteria were established in 2001 by a *Joint FAO/WHO expert consultation on foods derived from biotechnology* for assessing the risks that may be associated with the preparation of such foods. These include the genetic constructs used in the GM microorganisms, the pathogenic potential of GMM, and the detrimental effects of a potential gene transfer. For GMMs used in foods, the ensuing risk assessment ought to focus on the effects of a possible interaction between the GMM and endogenous intestinal microflora and the potential immune-stimulatory or immunomodulatory effects of the microorganisms in the event the gastrointestinal tract is colonized. Small regulatory elements derived from viral DNA are commonly used to drive the expression of transgenes in GMOs. Viral-DNA constructs are sometimes used as transgenes to establish resistance against viral pests, as they express viral proteins that confer viral resistance on plants. Some scientists suggest that the potential interaction of GM viral constructs with related wild-type viruses needs to be part of the risk assessment, to evaluate the potential of new viral pest strains evolving through mechanisms of recombination. Insertion of viral vectors into functionally important genes of recipient patients in the field of biomedicine has been reported, and whereas such vectors are not commonly used in food production, this evidence indicates the limited understanding of mechanisms directing insertion of genetic constructs.

Safety Aspects of Foods Derived from Biopharming

The potential to produce human proteins in animals has resulted in great interest in new possibilities for human health, but also in efforts to establish appropriate risk-assessment methods. The biosafety aspects of molecular 'farming' can be divided into two major groups: the potential spread of transgenes; and the potential negative effects of the expressed protein on the environment and the consumer. Practices and guidelines ensuring effective separation of 'biopharming' are being investigated. Experts agree that the risk assessment should ensure that proteins designed to produce pharmaceutical products, *e.g.* in the animal's milk, cannot find their way to other parts of the animal's body, possibly causing adverse effects.

Potential Effect of GMOs on Human Health Mediated Through Environmental Impact

Work on environmental-health indicators suggests that various agricultural practices have direct and indirect effects on human health and development. Hazards can take many forms—wholly natural in origin, or derived from human activities and interventions. The need to assess indirect effects of the use of GMOs in food production has been emphasized by many countries. Potential environmental-health hazards from the release of GMOs into the environment have been discussed in a report by WHO and the Italian Environment Protection Agency, in which health effects have been analysed "as an integrating index of ecological and social sustainability". For example, the production of chemicals or enzymes from contained GM microorganisms have contributed significantly to decreases in the amount of energy use, toxic and solid wastes in the environment, thereby significantly enhancing human health and development.

A further example of beneficial human/environmental outcomes of introducing GM crops is the reduction in the use, environmental contamination and human exposure to pesticides demonstrated in some areas. This has occurred especially through the use of pesticide-resistant *Bt* cotton, which has been shown to decrease pesticide poisoning in farm workers. Outcrossing of GM plants with conventional crops or wild relatives, as well as the contamination of conventional crops with GM material, can have an indirect effect on food safety and food security by contamination of genetic resources. Although initial concerns about introgression of transgenic DNA into traditional landraces of maize in Mexico arose as a result of the findings of transgenic DNA in such landraces in 2000 recently published results from samples taken during a broad, systematic survey in 2003 and 2004 in the same region shows no transgenes in these landraces. Still, the potential for introgression remains a possibility and risk-mitigation measures are being considered.

Both outcrossing and contamination characteristics are dependent on the pollination and distribution characteristics of pollen and seeds of the specific plant. In the USA, GM 'Starlink' maize was not approved for food use, but unintentionally started to appear in maize food products. This example demonstrated the problem of contamination and highlighted the potential for unintended impacts on human health and safety.

In the case of Starlink maize, full segregation of GM varieties not intended for food use and other varieties of the same crop species could not be achieved. Improved molecular methods for containment of the transgenes as well as farm management measures are under discussion, *e.g.* isolation distances, buffer zones, pollen barriers, control of volunteer plants, crop rotation and planting arrangements for different flowering periods, and monitoring during cultivation, harvest, storage, transport and processing. The likelihood of GM animals entering and persisting in the environment will vary among taxa, production systems, modified traits, and receiving environments.

The spread and persistence of GM fish and shellfish—or their transgenes—in the environment could be an indirect route of entry of GM animal products into the human food supply. This is because escaped individuals or their descendents could subsequently be captured in fishing for those species. Similar mechanisms might apply for poultry such as ducks and quail that are subject to sport or subsistence harvest. Live transport and sale of GM fish and poultry pose another route for the escape of GM animals and their entry into the environment.

GMOs and Environmental Safety

Principles of Environmental Risk Assessment

In many national regulations, the elements of environmental risk assessment for GM organisms include the biological and molecular characterizations of the genetic insert, the nature and environmental context of the recipient organism, the significance of new traits of the GMO for the environment, and information on the geographical and ecological characteristics of the environment in which the introduction will take place.

The risk assessment focuses especially on potential consequences for the stability and diversity of ecosystems, including putative invasiveness, vertical or horizontal gene flow, other ecological impacts, effects on biodiversity and the impact of the presence of GM material in other products. Different approaches in the ERA regulations of different countries have often resulted in different conclusions on the environmental safety of certain GMOs, especially where the ERA focuses not only on the direct effects of GMOs, but also addresses indirect or long-term effects on ecosystems, *e.g.* impact of agricultural practices on ecosystems. Internationally, the concept of 'familiarity' was developed also in the concept of environmental safety of transgenic plants.

The concept facilitates risk/safety assessments, because to be familiar means having enough information to be able to make a judgement of safety or risk. Familiarity can also be used to indicate appropriate management practices, including whether standard agricultural practices are adequate or whether other management practices are needed to manage the risk. Familiarity allows the risk assessor to draw on previous knowledge and experience with the introduction of plants and microorganisms into the environment and informs appropriate management practices.

As familiarity depends also on knowledge of the environment and its interaction with introduced organisms, the risk/safety assessment in one country may not be applicable in another country. Currently, the Cartagena Protocol on Biosafety of the Convention on Biological Diversity is the only international regulatory instrument which deals specifically with the potential adverse effects of GMOs on the environment, taking also into account effects on human health. The Protocol covers transboundary movements of any GM foods that meet the definition of an LMO. Annex III of the Protocol specifies general principles and methodology for risk assessment of LMOs.

The Protocol establishes a harmonized set of international rules and procedures designed to ensure that countries are provided with the relevant information, through the information exchange system called the Biosafety Clearing House. This Internet-based information system enables countries to make informed decisions before agreeing to the importation of LMOs. It also ensures that LMO shipments are accompanied by appropriate identification documentation. While the Protocol is the key basis for international regulation of LMOs, it does not deal specifically with GM foods, and its scope does not consider GM foods that do not meet the definition of an LMO. Furthermore, the scope of its consideration of human-health issues is limited, given that its primary focus is biodiversity, in line with the scope of the Convention itself. Consequently, the Protocol alone is not sufficient for the international regulation of GM foods.

Potential Unintended Effects of GMOs on Non-Target Organisms, Ecosystems and Biodiversity

This study does not focus specifically on the effects that GMOs used in food production may have on the environment. However, these aspects need to be considered in a holistic assessment of GM food production, as environmental effects may indirectly affect human health and development in many ways. Potential risks for the environment include unintended effects on non-target organisms, ecosystems and biodiversity. Insect-resistant GM crops have been developed by expression of a variety of insecticidal toxins from the bacterium *Bacillus thuringiensis*. Detrimental effects on beneficial insects, or a faster induction of resistant insects have been considered in the ERA of a number of insect-protected GM crops. Studies on the toxicity of *Bt* maize on the monarch butterfly in the USA indicate that for most commercially available hybrids, the *Bt* expression in pollen is low, and laboratory and field studies show that no acute toxic effects at any pollen density would be encountered in the field.

These questions are considered an issue for monitoring strategies and improved pest-resistance management. Increased doses of herbicide can be applied post-emergence to herbicide-tolerant crops, thus avoiding routine pre-emergence applications and reducing the number of applications needed. Also, the need for tilling can be reduced under critical soil conditions. In certain agro-ecological situations, such as a high weed pressure, the use of herbicide-tolerant crops has resulted in a reduction in quantity of the herbicides used. However, in other cases, herbicide use has stayed the same or even increased. In other situations, the following have been investigated: potentially detrimental consequences for plant biodiversity, weed shifts to less-sensitive species and development of herbicide resistance, decreased biomass, adverse effects on wildlife such as arthropods or birds, or consequences for agricultural practices, *e.g.* the use of the ecologically important practice of crop rotation.

Outcrossing

Outcrossing of transgenes has been reported from fields of commercially grown GM plants, including oilseed rape and sugar beet, and has been demonstrated in experimental releases for a number of crops, including rice and maize. Outcrossing could result in an undesired transfer of genes such as herbicide-resistance genes to non-target crops or weeds, creating new weed-management problems.

The consequences of outcrossing can be expected in regions where a GM crop has a sympatric distribution and synchronized flowering period that is highly compatible with a weedy or wild relative species, as demonstrated for rice. In view of the possible consequences of gene flow from GMOs, the use of molecular techniques to inhibit gene flow has been considered and is under development. Isolation distances or future molecular strategies for transgene confinement in transgenic crops may reduce gene flow. Stringent isolation measures may be necessary because of complex dispersal mechanisms for certain crops. Gene confinement techniques, *e.g.* introducing the transgene in plastids which are not inherited paternally, are either not very effective because of gene flow via seeds or they are still in an early stage of development.

GM Animals

The possibility that certain genetically engineered fish and other animals may escape, reproduce in the natural environment and introduce recombinant genes into wild populations is raised in a report of a United States Academy of Science study. Genetically engineered insects, shellfish, fish and other animals that can easily escape, are highly mobile and that form feral populations easily, are of concern, especially if they are more successful at reproduction than their natural counterparts. For example, it is possible that if released into the natural environment, transgenic salmon, with genes engineered to accelerate growth could compete more successfully for food and mates than wild salmon, thus endangering wild populations.

The use of sterile, all-female genetically engineered fish could reduce interbreeding between native populations and farmed populations, a current problem with the use of non-engineered fish in ocean net-pen farming. Sterility eliminates the potential for spread of transgenes in the environment, but does not eliminate all potential for ecological harm. Monosex triploidy is the best existing method for sterilizing fish and shellfish, although robust triploidy verification procedures are essential.

GMMs

Gene transfer from bacteria to bacteria in the soil has been demonstrated in some systems, *e.g.* for antibiotic-resistance genes and only a limited number of releases of GMMs has been permitted; mainly to explore the spread and the fate of microorganisms in nature. Risk assessment in this field is impeded by a number of factors, such as the limited knowledge of microorganisms in the environment, the existence of natural transfer mechanisms between microorganisms, and the difficulties in controlling their spread.

Status of Methods for Estimating Potential Environmental Entry

Methods by which to reliably characterize potential environmental entry have not yet been standardized. Net-fitness methodology does provide, however, a systematic and comprehensive approach based on contemporary evolutionary and population biology. It involves a two-step process of measuring fitness-component traits covering the entire life-cycle for GM animals, their conventional counterparts, and crosses between the two; and entering the fitness data from step 1 into a simulation model that predicts the fate of the transgene across multiple generations. There is a need to validate the predictions based on this method. Initial experiments are under way to this end.

Regional Specificity in Safety Assessments

Contradictory findings for benefits and risks for the same GM crop may reflect that such effects may be a consequence of different agro-ecological localities or regions. For example, the use of herbicide-resistant crops could potentially be detrimental in a small-sized agricultural area which has extensive crop rotation and low levels of pest pressure. Moderate herbicide uses on these GM plants could be beneficial in other agricultural situations. At present, no conclusive evidence on environmental advantages or costs can be generalized from the use of GM crops. Consequences may vary significantly between different GM traits, crop types and different local conditions including ecological and agro-ecological characteristics. In the USA, the overall difference in herbicide use between GM and conventional soybeans ranged from +7 to −40% with an average reduction of 10%. These changes have been associated with a number of factors including soil type, weed pressure, farm size, management style, prices of different herbicide programmes, and climate. Potential advantages of *Bt* maize have been widely attributed to regions with a significant pest pressure from the maize borer. The consequences of outcrossing can produce highly different characteristics, depending on the potentially different recipient plants in different ecological regions. These observations suggest that risk assessment needs to reflect the regional specificities of the receiving environments in addition to the characteristics of the GMO. In 1999, the government of the United Kingdom asked an independent consortium of researchers to investigate how growing GM crops might affect the abundance and diversity of farmland wildlife compared with growing conventional varieties of the same crops. In the largest-ever field trials of GM crops in the world, the researchers compared three GM crops with their conventional counterparts. The crops were sugar- and fodder-beet, spring-sown oilseed rape, and maize. The GM crops had been genetically modified to make them resistant to specific herbicides. Other types of GM crops, such as those engineered to be resistant to certain insect pests, were not included in the study.

The team found that there were differences in the abundance of wildlife between GM herbicide-tolerant crop fields and conventional crop fields. Growing conventional beet and spring rape was better for many groups of wildlife than growing GM herbicide-tolerant beet and spring rape. There were more insects, such as butterflies and bees, in and around the conventional crops because there were more weeds to provide food and cover. There were also more weed seeds in conventional beet and spring rape crops than in their GM counterparts. Such seeds are important in the diets of some animals, particularly some birds. In contrast, growing GM herbicide-tolerant maize was better for many groups of wildlife than conventional maize. There were more weeds in and around the GM herbicide-tolerant crops, more butterflies and bees at certain times of the year, and more weed seeds. The researchers stress that the differences they found do not arise just because the crops have been genetically modified. They arise because these GM crops give farmers new options for weed control. That is, they use different herbicides and apply them differently.

The results of this study suggest that growing such GM crops could have implications for wider farmland biodiversity. However, other issues will affect the medium- and long-term impacts, such as the areas and distribution of land involved, how the land is cultivated and how crop rotations are managed. These make it hard for researchers to predict the medium- and large-scale effects of GM cropping with any certainty. In addition, other management decisions taken by farmers growing conventional crops will continue to have an impact on wildlife. Monitoring of the environmental impacts of GM crops in various regions and from investigation over longer time periods may be necessary to conclude on effects and consequences.

Monitoring of Human Health and Environmental Safety

In the future, GMOs may gain wider approval for environmental release, either with or without approval to enter them in the human food supply. In such situations, it will be important to consider whether or not to apply postmarket monitoring for unexpected environmental spread of the GMOs and their transgenes) that may pose food safety hazards.

Methods for detection of such GMOs and their transgenes in the environment are likely to involve application of two well-established bodies of scientific methodologies:

1. Diagnostic, DNA-based markers; and
2. Sampling protocols that are adequate and cost-effective.

However, there is a need to fully develop appropriate protocols for application of these methods to postmarket detection of environmental spread of GMOs and their transgenes. Monitoring can also be helpful to assure confinement of GMOs during R&D. Postmarket monitoring of GM foods with respect to direct human-health impacts has been raised at international conferences and within the Codex Alimentarius Commission. Opinions regarding such monitoring vary from neither necessary nor feasible, to being essential to support and improve the results of a risk assessment and enable an early detection of uncharacterized and unintended hazards. Some have suggested that monitoring of potential long-term effects of GM foods with significantly altered nutritional composition should be mandatory.

The Expert consultation on the safety assessment of foods derived from GM animals held in 2003 identified a need for postmarket surveillance, and therefore a product-tracing system, for:

- Confirmation of the assessments made during the premarket phase;
- Assessment of allergenicity or long-term effects; and
- Unexpected effects.

The issue of postmarket surveillance is closely related to risk characterization. In general, potential safety issues should be addressed adequately through premarket studies, as the potential of postmarket studies is currently very limited. Postmarket surveillance could be useful

Food Applications for Human Health

in certain instances where clear-cut questions require, for instance, a better estimate of dietary exposure and/or the nutritional consequences of GMO-derived food. Tools to identity or trace GMOs or products derived from GMOs in the environment or food-chain are a prerequisite for any kind of monitoring. Detection techniques are in place in a number of countries to monitor the presence of GMOs in food, to enable the enforcement of GM labelling requirements, and to monitor effects on the environment. Attempts to standardize analytical methods for tracing GMOs have been initiated.

Conclusions

GM foods currently available on the international market have undergone risk assessments and are not likely to present risks for human health any more than their conventional counterparts. The risk-assessment guidelines specified by CAC are thought to be adequate for the safety assessment of GM foods currently on the international market. Guidelines for environmental risk assessment have been developed under the Convention on Biological Diversity. The potential risks associated with GMOs and GM foods should be assessed on a case-by-case basis, taking into account the characteristics of the GMO or the GM food and possible differences of the receiving environments. In the field of potential risks derived from outcrossing or contamination from GM crops, relevant consequences need to be investigated for specific crops, and strategies for risk management need to be explored. As highlighted in the Codex *Principles for the risk analysis of foods derived from modern biotechnology*, the assessment of the potential of GM foods to elicit hypersensitivity reactions should be part of the risk assessment for GM foods.

This includes a general analysis of the proteins expressed and assessment of the specific properties of the GM food under consideration to elicit hypersensitivity reactions. A better understanding of the impact and interaction of food with the immune system is required to decipher how and whether conventional and GM foods cause specific health and safety problems. New methodology for the development of GMOs may significantly reduce potential risks derived from the random integration of transgenes used in current methods.

8

Chemical Hazards Associated with the Food Supply

Changes in our society and our food supply have raised new concerns about food chemical safety. Some of the changes that have raised new concerns about foodborne infectious disease are also affecting how government agencies carry out the more familiar task of protecting the food supply from toxic chemical agents.

Foods are themselves a complex collection of naturally occurring chemicals that have nutritive, organoleptic, and pharmacological functions and occasionally toxic effects. Naturally occurring toxicants probably present risks second only to those imposed by microorganisms. These natural toxicants include seafood toxins and foodborne mycotoxins.

Other chemicals are introduced into foods intentionally or unintentionally. Intentionally added substances include food and colour additives, flavours, enzymes, vitamins and minerals, and other ingredients that help to add value or characteristic properties or functions to a food. Because of broad public concern about synthetic chemicals, the toxicologic profiles of many of these synthetic components of food are much more complete than those of natural components of food. As a result, the risks posed by regulated food additives are generally better characterized than those of many naturally occurring substances. Unintentional additives can include environmental or industrial contaminants as well as some substances used in food production but not intended to be part of the food. The migration of food production substances into the final product is generally very low, but must be carefully regulated to ensure safety. Examples include sanitizers used to keep food production surfaces safe, packaging materials used to keep food safe and fresh, pesticides used on crops and drugs used in animals to mitigate damage, disease, microbial toxin production, and general food losses. Environmental or industrial contaminants are not sanctioned but have the potential to enter the food chain. Examples of chemical contamination incidents are methyl mercury in fish, mistaken mixing of polybrominated biphenyls (a fire retardant) into animal feed, and leakage of ammonia refrigerant into frozen foods.

Some people are sensitive or allergic to chemical constituents that are harmless to the rest of the population. These allergic reactions are estimated to occur in 1 to 2 percent of adults and 5 to 8 percent of children. Although serious reactions are rare, it is estimated that several dozen deaths occur each year because of allergic reactions to food. More than 160 foods and food-related substances have been identified as being able to cause allergic reactions. However, in the United States, more than 90 percent of allergic reactions appear to be caused by just eight food types: peanuts, tree nuts (such as walnuts, pecans, almonds, hazel nuts), crustacea (shrimp, lobster, and other shellfish), eggs, milk, soy, fish, and wheat.

Consequently, the subject of chemical hazards in food is complex and includes consideration of potential risks that vary widely in scope and severity. Many chemical hazards associated with foods have been recognized only in the last century as advancements have been made in chemistry, toxicology, and risk assessment. Public concern over these hazards has grown in recent decades, in line with the increasing distrust of chemicals generally and of their use in the environment. Indeed, there have been episodes of chemical intoxication (usually arising out of accident or occupational exposure) with tragic consequences. Although most experts agree that the more serious hazards in the American food supply are not chemical but microbiological, public concern has demanded that proportionally more regulatory resources be applied to chemical hazards.

New Food Components

The American public's growing interest in the relationship between diet and health has led to an increased demand for foods or food constituents that not only have nutritive value but also hold promise for prevention or even treatment of disease. These products have been referred to as dietary supplements, functional foods, pharma-foods or nutraceuticals. In a recent study, more than two-thirds of surveyed households reported use of a vitamin, mineral, or herbal supplement within the previous six-month period. The Dietary Supplement Health and Education Act of 1994 eased restrictions on certain statements of nutrition support made for supplements and exempted them from the safety approval requirement applicable to conventional food additives. That legal change helped spur the growing market for supplements of all types, including herbal products, and raises food safety concerns. Some supplements and herbal products on the market may pose a risk of adverse health effects because they are not required to meet specified safety standards before being sold. They may thus contain varying amounts or unknown or inadequately characterized ingredients that can have pharmacological activity that has not been adequately characterized.

Food processors are examining relatively new sources of ingredients for more conventional functional properties. For example, gums and fibres such as konjac flour, tara gum, inulin, or psyllium fibre may provide bulk and texture to foods, yet have had limited food use in the United States. Potential broad use of these ingredients in domestic foods raises important safety questions: are there significant intestinal effects such as blockage or reduced transit times? Are there impacts on vitamin uptake from the intestinal tract or on osmotic balance? What are the potential allergy risks from increased exposure to plant-sourced materials? These are a few of the safety questions that may be raised when a new ingredient source is considered.

Food processors are also utilizing macronutrient substitutes, such as non-nutritive sweeteners and fat replacers in many food products. These macronutrient substitutes have potential value by lowering calories, sugar, or fat in food products. Because dietary quantities of these substitutes could be substantially higher than the amounts of typical food additives, the assessment of their safety can be particularly important as well as particularly difficult.

New Food Technologies

Modification of plants or animals via genetic engineering can improve yields and increase resistance to pests. This new technology might offer improvements in food safety through increased resistance to molds that produce food mycotoxins or through lower levels of allergenic proteins, fatty acids, or other undesirable components of food. However, there are important differences between countries in how food products from genetically modified organisms are regulated. Several products derived from genetic engineering have been declared safe by US regulators, but many European countries have either forbidden their sale or insisted on what marketers believe is disparaging labelling. Concerns about the safety of products from genetically engineered plants and animals are only partially resolved.

Food irradiation is not a new technology. Irradiation of fruits and vegetables decreases the risk of pathogens and extends shelf life. But as discussed, public concerns about the safety of irradiated food, fuelled in part by a lack of consumer education, limit its use in the United States. These concerns prompted the requirement that certain irradiated foods be labelled as such. The extent to which this labelling requirement has limited subsequent adoption of the process is unknown.

Re-emerging Toxic Agents

As the science of toxicology has progressed over the last 50 years, questions about the safety of chemicals in foods have become more sophisticated. When tougher safety standards are applied to new food chemicals or reapplied to old ones, new issues of toxicity emerge. The cycle of re-evaluating safety standards began in the years after World War II when advances in the understanding of the mechanisms of carcinogenesis coincided with the increased use of rodent bioassays. The newly focused attention to carcinogens that ensued led Congress to pass the 1958 "Delaney clause" proscription of carcinogens. The focus on carcinogens in food continues to consume substantial testing and regulatory resources. Further advances in biomedical research have raised new concerns and new standards of safety that now address teratogenicity, reproductive toxicity, mutagenicity, hormonal effects, and immunotoxicity.

The effects of food constituents on hormonal function is of concern as medical research has indicated the importance of certain hormones in regulating diseases such as breast cancer and osteoporosis. This phenomenon, sometimes referred to as endocrine disruption, led Congress in 1996 to mandate special regulatory attention to the issue and inspired the Environmental Protection Agency (EPA) to convene an expert panel, the Endocrine Disrupter Screening and Testing Advisory Committee. That committee has attempted to define endocrine disrupters, determine analytical methods, and recommend how the disrupters should be regulated. Potential candidates for evaluation include constituents of some food packaging materials, pesticides, and natural food constituents, such as genestein in soy.

Some groups have recently asked whether chemical safety assessments provide adequate safety margins to protect children. That concern draws attention because new pesticide legislation directs EPA to consider the aggregate risks posed by any pest control agent; that is, exposures should be assessed for all potential sources of a chemical and for other chemicals with a similar mechanism of action. Another concern is the emergence of evidence of subtle developmental and behavioural effects of relatively low concentrations of chemicals (for example, lead).

In summary, chemical hazards, as well as foodborne pathogens, present new and changing challenges to the food safety system. Federal food safety efforts will need full integration and sufficient support to meet these challenges.

Physical Hazards

The foods we consume begin as raw agricultural commodities grown in open fields or waters or raised in a variety of production facilities, such as barns, coops, pens, and feedlots. Rocks, stones, metal, wood, glass, and other physical objects can become part of raw ingredients. Further contamination can occur in the transport, processing, or distribution of foods because of equipment failure, accident, or negligence.

Foreign physical materials in foods can cause serious harm to consumers. Protective devices that remove or prevent physical hazards include metal detectors, magnets, sieves, traps, scalpers, and screens. Other effective means of protection are production plant policies against the use of glass, wood, or non-ferrous metal where possible; employee training; quality audits of ingredient suppliers; and sensory tests.

Federal agencies have established "defect action levels" for natural or unavoidable defects that do not affect human health, such as stems or pits in fruits and vegetables, bone in mechanically deboned meats, and microscopic insect fragments, which are primarily of aesthetic concern and do not present a safety hazard.

The Changing Nature of Food Hazards

- Changes in the risk of infectious foodborne disease are due primarily to:
- Changes in diet,
- Increasing use of commercial food service and in food eaten or prepared away from home,
- New methods of producing and distributing food,
- New or re-emerging infectious foodborne agents, and

- The growing number of people at high risk for foodborne illnesses due to:
  - Increasing number of elderly, and
  - Increasing number of people with depressed immunity or resistance to infection.
- Changes in chemical hazards associated with the food supply must be monitored and evaluated; these include:
- Increased use of dietary supplements and herbal products without requirements to meet specified safety standards;
- New food components that mimic attributes of traditional food components;
- Introduction of new food technologies and processes;
- Changes in presence of food toxins and additives, including unintentional food additives; and
- Presence of physical hazards associated with new technologies or sources of foods.

CAUSE FOR INCREASING CONCERN

There have been dramatic changes in the Indian food supply. These changes have contributed to recent outbreaks of infectious foodborne illness, which in turn led to the request for this committee to examine aspects of the Indian food safety system. The committee recognizes the growing concern for controlling the microbiological hazards related to food, but believes that government attention must also be addressed to chemical and physical hazards. This chapter is organized in two parts: the first describes major changes that affect the epidemiology of infectious foodborne disease, and the second describes examples of potential chemical hazards which have emerged in part from some of the same changes in the food supply.

The epidemiology of foodborne disease

Recent outbreaks of foodborne disease caused by many different pathogens and involving a variety of food products have been the subject of headlines, but it is unclear whether the incidence of foodborne disease has increased over the last generation. The major reason for the uncertainty is that the lack of a national foodborne-disease surveillance system has prevented the study of trends in disease rates. What is clear is that factors affecting the potential safety of the nation's food supply have changed dramatically over the last generation and justify concern that the incidence of foodborne disease is high and may be increasing.

At least five trends contribute to the possible increase in foodborne disease: changes in diet, the increasing use of commercial food services, new methods of producing and distributing food, new or re-emerging infectious foodborne agents, and the growing number of people at high risk for severe or fatal foodborne diseases.

Diet

Annual food expenditures in the United States, as a share of disposable personal income, decreased from 14 percent to 11 percent from 1970 to 1996. No industrialized nation spends a smaller share of its wealth on food than the United States. For much of the population, readily available food is more varied and more affordable than ever before. For example, in the 1960s, an average US grocery store had fewer than 7,000 food items available. Today, an average US grocery store sells about 30,000 food items, and over 12,000 new products are introduced each year.

During this time when relative costs of the US food supply are decreasing, per capita consumption of many foods has changed substantially. Public health efforts to promote a "heart-healthy" diet have helped to boost the consumption of fresh fruits and vegetables. On a per capita basis, in 1995, Americans ate about 31 lb more commercially grown vegetables, including potatoes and sweet potatoes, and 24 lb more fresh fruit than in 1970. As the consumption and variety of produce have increased, so has the importation of produce from developing countries. The General Accounting Office estimates that in 1995 one-third of all fresh fruit consumed in the United States was imported. Food imports have increased both because of lower production costs in foreign countries and because of consumer demand for year-round supplies of fruits and vegetables that have limited growing seasons in the United States.

For example, 17 percent of cantaloupes, 52 percent of green onions, 36 percent of cucumbers, and 34 percent of tomatoes sold in the United States in 1996 were grown in Mexico. Seasonally, as much as 79 percent of a particular commodity consumed in this country has been raised in Mexico alone, and the percentage of produce from other developing countries consumed here is growing rapidly.

Fresh produce items were the leading vehicle associated with foodborne disease outbreaks in Minnesota from 1990 to 1996, accounting for almost one-third of all outbreaks. This percentage is higher than that available from national foodborne disease surveillance data and possibly reflects more active surveillance in that state. Other trends in the United States are the decreasing consumption of beef and the increased consumption of chicken and seafood. In 1970, the average American consumed 79 lb of beef, 27 lb of chicken, and 12 lb of fish and shellfish; in 1996, annual per capita consumptions were 64, 50 and 15 lb, respectively. Contamination of red meat with *Salmonella and Escherichia coli* 0157:H7 remains important, but the risk posed by chicken as a vehicle for *Campylobacter and Salmonella* has grown substantially.

Cultural changes affect not only what Americans eat, but also where they eat and how their food is prepared. Increasingly, Americans have time-pressured lifestyles. Saving time and effort in shopping for and preparing food will continue to be important for many Americans. Households with a single parent or two working adults often face particular pressures in food shopping and meal preparation that can affect food selection and safety.

Cookbook sales and television cooking shows demonstrate that cooking is popular, but there seems to be decreased interest in ordinary daily home food preparation and, with most adults working, a lack of role models in the kitchen. Reductions in time for and interest in home food preparation also result in changed food patterns, fewer homemade dishes, more reliance on leftovers, increased purchase of prepared or convenience foods, and frequent eating away from home.

With less time spent in the kitchen and greater availability of high-quality, ready-to-eat dishes and convenience items, Americans' food preparation skills are diminishing. As a result, appreciation of simple but critical food safety techniques, such as washing hands and utensils and storing foods at optimal temperatures, has likely diminished.

Individual food tastes and preparation styles are brought to the United States from around the world, and the increasing ethnic diversity of the American population may affect food safety in several ways.

Different ethnic groups have different concerns and practices regarding food safety, and this could affect the activities of immigrants as food preparers at home and in the workplace. New food risks can arise as immigrant populations adapt traditional preferences and practices to their new environment. Food safety education programmes should consider the food beliefs and practices of various cultural groups.

Although data show a rise in perceived risk of foodborne illness among consumers, attitudes do not always translate into improved food-handling practices. Over half of all shoppers report washing their hands and/or food preparation surfaces, yet only 28 percent know that cooking temperatures are critical and that foods should be refrigerated promptly. A recent Food and Drug Administration study found that many US consumers still eat undercooked hamburger meat and raw eggs.

Commercial Food Services

The percentage of the food dollar spent on food consumed away from home has risen dramatically over the last three decades. In 1970, only 34 percent of our food dollars were spent eating away from home. In contrast, in 1996, 46 percent of our food dollars were spent for meals and snacks prepared outside the home. Institutional feeding sites serve a wide range of people—from very young children in child care centres to the elderly—in congregate sites for meals, alternative-care centres, and nursing homes. Meals are also prepared as takeout foods from supermarkets and convenience stores; many of these meals include one or more cold food items, such as delicatessen sandwiches and salads that require extensive food handling and are not cooked before consumption. These changes have led to an increase in the number of people handling food and the potential for an increase in the transmission of foodborne diseases from food handlers to consumers.

The average food handler in this country earns the minimum wage, lacks sick leave and other health benefits, and has very limited opportunities for advancement. These jobs are filled by people with few employment opportunities and low economic status, conditions that may be related to a high incidence of intestinal diseases and low rates of routine hand washing. Recent evidence from Minnesota demonstrates the increased risk of food handler associated transmission of *Salmonella typhimurium;* this finding was documented because *Salmonella* isolates in Minnesota undergo molecular characterization and epidemiologic investigation. Food handlers in other areas of the United States probably play a similar role in the transmission of *Salmonella* and other enteric agents.

Production and Distribution

The changing availability and sources of our food supply have brought changes in methods of producing and distributing food. Today, large manufacturing plants can process quantities and types of products that two decades ago would have required many smaller plants. Although the ability to control possible hazards increases when there are fewer plants, the potential for larger outbreaks—even if product contamination is minimal—is evident. Thousands of people become ill during such outbreaks. For example, contamination of tanker trucks with nonpasteurized liquid eggs and subsequent use of those trucks to haul pasteurized ice cream mix most likely led to the largest documented outbreak of salmonellosis in the United States. That outbreak showed that even sporadic low level contamination of a single product can result in a major epidemic of foodborne illness because of the quantity of product consumed.

In the spring of 1998, a cereal produced by the largest generic label manufacturer of cereal in the United States caused a national outbreak of *Salmonella agona* infection. Outbreaks such as this can be difficult to detect because contamination of the product is sporadic and the product is marketed widely.

The driving forces of globalization, advanced technology, and economic competitiveness have dramatically affected the structure and practices of livestock and poultry production. Herds, flocks, and other populations of food animals, including fish and seafood, are increasingly concentrated in fewer and larger production units. The traditional farmstead model of the past, often characterized by multiple species and small numbers of food animals reared on a single farm, has been replaced by specialized, large-scale production systems. For example, from 1994 to 1995, the number of US hog operations decreased from 207,980 to 182,700, but both the inventory of hogs and the total number of hogs produced increased during the same time. That mirrors the general agribusiness trend, and the restructuring continues.

9

Foodborne Pathogens

Foodborne pathogens are the leading causes of illness and death in less developed countries, killing approximately 1.8 million people annually. In developed countries, foodborne pathogens are responsible for millions of cases of infectious gastrointestinal diseases each year, costing billions of dollars in medical care and lost productivity. New foodborne pathogens and foodborne diseases are likely to emerge, driven by factors such as pathogen evolution, changes in agricultural and food manufacturing practices, and changes to the human host status. There are growing concerns that terrorists could use pathogens to contaminate food and water supplies in attempts to incapacitate thousands of people and disrupt economic growth.

Enteric Viruses

Food and waterborne viruses contribute to a substantial number of illnesses throughout the world. Among those most commonly known are hepatitis A virus, rotavirus, astrovirus, enteric adenovirus, hepatitis E virus, and the human caliciviruses consisting of the noroviruses and the Sapporo viruses. This diverse group is transmitted by the fecal-oral route, often by ingestion of contaminated food and water.

Protozoan Parasites

Protozoan parasites associated with food and water can cause illness in humans. Although parasites are more commonly found in developing countries, developed countries have also experienced several foodborne outbreaks. Contaminants may be inadvertently introduced to the foods by inadequate handling practices, either on the farm or during processing of foods. Protozoan parasites can be found worldwide, either infecting wild animals or in water and contaminating crops grown for human consumption. The disease can be much more severe and prolonged in immunocompromised individuals.

Mycotoxins

Molds produce mycotoxins, which are secondary metabolites that can cause acute or chronic diseases in humans when ingested from contaminated foods. Potential diseases include cancers and tumors in different organs (heart, liver, kidney, nerves), gastrointestinal disturbances, alteration of the immune system, and reproductive problems.

Species of Aspergillus, Fusarium, Penicillium, and Claviceps grow in agricultural commodities or foods and produce the mycotoxins such as aflatoxins, deoxynivalenol, ochratoxin A, fumonisins, ergot alkaloids, T-2 toxin, and zearalenone and other minor mycotoxins such as cyclopiazonic acid and patulin. Mycotoxins occur mainly in cereal grains (barley, maize, rye, wheat), coffee, dairy products, fruits, nuts and spices.

Control of mycotoxins in foods has focused on minimizing mycotoxin production in the field, during storage or destruction once produced. Monitoring foods for mycotoxins is important to manage strategies such as regulations and guidelines, which are used by 77 countries, and for developing exposure assessments essential for accurate risk characterization. Aflatoxins are still recognized as the most important mycotoxins.

They are synthesized by only a few Aspergillus species, of which A. flavus and A. parasiticus are the most problematic. The expression of aflatoxin-related diseases is influenced by factors such as age, nutrition, sex, species and the possibility of concurrent exposure to other toxins. The main target organ in mammals is the liver, so aflatoxicosis is primarily a hepatic disease. Conditions increasing the likelihood of aflatoxicosis in humans include limited availability of food, environmental conditions that favour mold growth on foodstuffs, and lack of regulatory systems for aflatoxin monitoring and control.

Yersinia Enterocolitica

Yersinia enterocolitica includes pathogens and environmental strains that are ubiquitous in terrestrial and fresh water ecosystems. Evidence from large outbreaks of yersiniosis and from epidemiological studies of sporadic cases has shown that Y. enterocolitica is a foodborne pathogen. Pork is often implicated as the source of infection. The pig is the only animal consumed by man that regularly harbors pathogenic Y.

enterocolitica. An important property of the bacterium is its ability to multiply at temperatures near 0°C, and therefore in many chilled foods. The pathogenic serovars (mainly O:3, O:5, 27, O:8 and O:9) show different geographical distribution. However, the appearance of strains of serovars O:3 and O:9 in Europe, Japan in the 1970s, and in North America by the end of the 1980s, is an example of a global pandemic. There is a possible risk of reactive arthritis following infection with Y. enterocolitica.

Vibrio

Vibrio species are prevalent in estuarine and marine environments, and seven species can cause foodborne infections associated with seafood. Vibrio cholerae O1 and O139 serovtypes produce cholera toxin and are agents of cholera. However, fecal-oral route infections in the terrestrial environment are responsible for epidemic cholera. V. cholerae non-O1/O139 strains may cause gastroenteritis through production of known toxins or unknown mechanism.

Vibrio parahaemolytitcus strains capable of producing thermostable direct hemolysin (TDH) and/or TDH-related hemolysin are most important causes of gastroenteritis associated with seafood consumption. Vibrio vulnificus is responsible for seafoodborne primary septicemia, and its infectivity depends primarily on the risk factors of the host.

V. vulnificus infection has the highest case fatality rate (50%) of any foodborne pathogen. Four other species (V. mimicus, V. hollisae, V. fluvialis, and V. furnissii) can cause gastroenteritis. Some strains of these species produce known toxins, but the pathogenic mechanism is largely not understood. The ecology of and detection and control methods for all seafoodborne Vibrio pathogens are essentially similar.

Staphylococcus Aureus

Staphylococcus aureus is a common cause of bacterial foodborne disease worldwide. Symptoms include vomiting and diarrhea that occur shortly after ingestion of S. aureus toxin-contaminated food. The symptoms arise from ingestion of preformed enterotoxin, which accounts for the short incubation time. Staphylococcal enterotoxins are superantigens and, as such, have adverse effects on the immune system.

The enterotoxin genes are accessory genetic elements in S. aureus, meaning not all strains of this organism are enterotoxin-producing. The enterotoxin genes are found on prophages, plasmids, and pathogenicity islands in different strains of S. aureus. Expression of the enterotoxin genes is often under the control of global virulence gene regulatory systems.

Campylobacter

Campylobacter spp., primarily C. jejuni subsp. jejuni is one of the major causes of bacterial gastroenteritis in the U.S. and worldwide. Campylobacter infection is primarily a foodborne illness, usually without complications; however, serious sequelae, such as Guillain-Barre Syndrome, occur in a small subset of infected patients. Detection of C. jejuni in clinical samples is readily accomplished by culture and nonculture methods.

Listeria Monocytogenes

Listeria monocytogenes is Gram-positive foodborne bacterial pathogen and the causative agent of human listeriosis. Listeria infections are acquired primarily through the consumption of contaminated foods, including soft cheese, raw milk, deli salads, and ready-to-eat foods such as luncheon meats and frankfurters.

Although L. monocytogenes infection is usually limited to individuals that are immunocompromised, the high mortality rate associated with human listeriosis makes it the leading cause of death among foodborne bacterial pathogens. As a result, tremendous effort has been made to develop methods for the isolation, detection and control of L. monocytogenes in foods.

Salmonella

Salmonella serotypes continue to be a prominent threat to food safety worldwide. Infections are commonly acquired by animal to human transmission though consumption of undercooked food products derived from livestock or domestic fowl. The second half of the 20th century saw the emergence of Salmonella serotypes that became associated with new food sources (*i.e.* chicken eggs) and the emergence of Salmonella serotypes with resistance against multiple antibiotics.

Shigella

Shigella species are members of the family Enterobacteriaceae and are Gram negative, nonmotile rods. Four subgroups exist based on O-antigen structure and biochemical properties: S. dysenteriae (subgroup A), S. flexneri (subgroup B), S. boydii (subgroup C) and S. sonnei (subgroup

D). Symptoms include mild to severe diarrhea with or without blood, fever, tenesmus and abdominal pain. Further complications of the disease may be seizures, toxic megacolon, reactive arthritis and hemolytic uremic syndrome. Transmission of the pathogen is by the fecal-oral route, commonly through food and water.

The infectious dose ranges from 10-100 organisms. Shigella spp. have a sophisticated pathogenic mechanism to invade colonic epithelial cells of the host, man and higher primates, and the ability to multiply intracellularly and spread from cell to adjacent cell via actin polymerization. Shigella spp. are one of the leading causes of bacterial foodborne illnesses and can spread quickly within a population.

Escherichia Coli

More information is available concerning Escherichia coli than any other organism, thus making E. coli the most thoroughly studied species in the microbial world. For many years, E. coli was considered a commensal of human and animal intestinal tracts with low virulence potential. It is now known that many strains of E. coli act as pathogens, inducing serious gastrointestinal diseases and even death in humans.

There are six major categories of E. coli strains that cause enteric diseases in humans, including the:
- Enterohemorrhagic E. coli, which cause hemorrhagic colitis and hemolytic uremic syndrome,
- Enterotoxigenic E. coli, which induce traveller's diarrhea,
- Enteropathogenic E. coli, which cause a persistent diarrhea in children living in developing countries,
- Enteroaggregative E. coli, which provokes diarrhea in children,
- Enteroinvasive E. coli that are biochemically and genetically related to Shigella species and can induce diarrhea,
- Diffusely adherent E. coli, which cause diarrhea and are distinguished by a characteristic type of adherence to mammalian cells.

Clostridium Botulinum and Clostridium Perfringens

Clostridium botulinum produces extremely potent neurotoxins that result in the severe neuroparalytic disease, botulism. The enterotoxin produced by C. perfringens during sporulation of vegetative cells in the host intestine results in debilitating acute diarrhea and abdominal pain.

Sales of refrigerated, processed foods of extended durability including sous-vide foods, chilled ready-to-eat meals, and cook-chill foods have increased over recent years. Anaerobic spore-formers have been identified as the primary microbiological concerns in these foods. Heightened awareness over intentional food source tampering with botulinum neurotoxin has arisen with respect to genes encoding the toxins that are capable of transfer to nontoxigenic clostridia.

Bacillus Cereus

The Bacillus cereus group comprises six members: B. anthracis, B. cereus, B. mycoides, B. pseudomycoides, B. thuringiensis and B. weihenstephanensis. These species are closely related and should be placed within one species, except for B. anthracis that possesses specific large virulence plasmids. B. cereus is a normal soil inhabitant, and is frequently isolated from a variety of foods, including vegetables, dairy products and meat. It causes a vomiting or diarrhea illness that is becoming increasingly important in the industrialized world. Some patients may experience both types of illness simultaneously. The diarrheal type of illness is most prevalent in the western hemisphere, whereas the emetic type is most prevalent in Japan. Desserts, meat dishes, and dairy products are the foods most frequently associated with diarrheal illness, whereas rice and pasta are the most common vehicles of emetic illness.

The emetic toxin (cereulide) has been isolated and characterized; it is a small ring peptide synthesised nonribosomally by a peptide synthetase. Three types of B. cereus enterotoxins involved in foodborne outbreaks have been identified. Two of these enterotoxins are three-component proteins and are related, while the last is a one-component protein (CytK). Deaths have been recorded both by strains that produce the emetic toxin and by a strain producing only CytK.

Some strains of the B. cereus group are able to grow at refrigeration temperatures. These variants raise concern about the safety of cooked, refrigerated foods with an extended shelf life. B. cereus spores adhere to many surfaces and survive normal washing and disinfection (except for hypochlorite and UVC) procedures. B. cereus food borne illness is likely under-reported because of its relatively mild symptoms, which are of short duration.

Food authenticity

It is important to be able to detect microorganisms in food, in particular pathogenic microorganisms or genetically modified microorganisms. Real-time PCR is an accepted analytical tool within the food industry. Its principal role has been one of assisting the legislative authorities, major manufacturers and retailers to confirm the authenticity of foods. The most obvious role is the detection of genetically modified organisms, but real-time PCR makes a significant contribution to other areas of the food industry, including food safety.

COMMON FOODBORNE PATHOGENS

Even though the United States has one of the safest food supplies in the world, there are still millions of cases of foodborne illness each year. Here are common foodborne pathogens (disease-causing microorganisms) with research-based information that includes:

- Cause of illness
- Incubation period
- Symptoms
- Possible contaminants
- Steps for prevention

Bacillus Cereus

- *Cause of illness:* Large molecular weight protein (diarrheal type) or highly heat-stable toxin (emetic type)
- *Incubation period:* 30 minutes to 15 hours
- *Symptoms:* Diarrhea, abdominal cramps, nausea, and vomiting (emetic type)
- *Possible contaminants:* Meats, milk, vegetables, fish, rice, potatoes, pasta, and cheese
- *Steps for prevention:* Pay careful attention to food preparation and cooking guidelines.

Campylobacter Jejuni

- *Cause of Illness:* Infection, even with low numbers
- *Incubation Period:* One to seven days
- *Symptoms:* Nausea, abdominal cramps, diarrhea, headache - varying in severity
- *Possible Contaminant:* Raw milk, eggs, poultry, raw beef, cake icing, water

- *Steps for Prevention:* Pasteurize milk; cook foods properly; prevent cross-contamination.

Clostridium Botulinum

- *Cause of Illness:* Toxin produced by Clostridium botulinum
- *Incubation Period:* 12 to 36 hours
- *Symptoms:* Nausea, vomiting, diarrhea, fatigue, headache, dry mouth, double vision, muscle paralysis, respiratory failure
- *Possible Contaminant:* Low-acid canned foods, meats, sausage, fish
- *Steps for Prevention:* Properly can foods following recommended procedures; cook foods properly.

Clostridium Perfringens

- *Cause of illness:* Undercooked meats and gravies
- *Incubation period:* 8 to 22 hours
- *Symptoms:* Abdominal cramps and diarrhea, some include dehydration
- *Possible contaminants:* Meats and gravies
- *Steps for prevention:* Proper attention to cooking temperatures.

Cryptosporidium Parvum

- *Cause of Illness:* Drinking contaminated water; eating raw or undercooked food; putting something in the mouth that has been contaminated with the stool of an infected person or animal; direct contact with the droppings of infected animals.
- *Incubation Period:* Two to 10 days
- *Symptoms:* Watery diarrhea accompanied by mild stomach cramping, nausea, loss of appetite. Symptoms may last 10 to 15 days.
- *Possible Contaminants:* Contaminated water or milk, person-to-person transmission (especially in child daycare settings). Contaminated food can also cause infections.
- *Steps for Prevention:* Avoid water or food that may be contaminated; wash hands after using the toilet and before handling food. If you work in a child care center where you change diapers, be sure to wash hands thoroughly with soap and warm water after every diaper change, even if you wear gloves. During communitywide outbreaks caused by contaminated drinking water, boil drinking

water for 1 minute to kill the Cryptosporidium parasite. Allow water to cool before drinking it.

Escherichia coli 0157:H7

- *Cause of Illness:* Strain of enteropathic E.coli
- *Incubation Period:* Two to four days
- *Symptoms:* Hemorrhagic colitis, possibly hemolytic uremic syndrome
- *Possible Contaminant:* Ground beef, raw milk
- *Steps for Prevention:* Thoroughly cook meat; no cross-contamination.

Giardia Lamblia

- *Cause of Illness:* Strain of Giardia lamblia Incubation Period: One to two weeks
- *Symptoms:* Infection of the small intestine, diarrhea, loose or watery stool, stomach cramps, and lactose intolerance.
- *Possible Contaminant:* Giardia is found in soil, food, water, or surfaces that have been contaminated with the feces from infected humans or animals..
- *Steps for Prevention:* Avoid swallowing contaminated recreational water (pools, hot tubs, fountains, lakes, rivers, ponds) or contaminated bathroom fixtures, toys, changing tables, diaper pails; avoid eating uncooked contaminated food; boil water for 1 minute before use or use a water filter with an absolute pore size of at least 1 micron or rated for "cyst removal." Cholorination or iodination of water may be less effective. Avoid fecal exposure during sexual activity

Hepatitis A

- *Cause of illness:* Hepatitis A Virus (HAV)
- Incubation period:
- *Symptoms:* Fever, malaise, nausea, abdominal discomfort
- *Possible contaminants:* Water, fruits, vegetables, iced drinks, shellfish, and salads
- *Steps for prevention:* Carefully wash hands with soap and water after using a restroom, changing a diaper, and before preparing food.

Listeria Monocytogenes

- *Cause of Illness:* Infection with Listeria monocytogenes
- *Incubation Period:* Two days to three weeks
- *Symptoms:* Meningitis, septicemia, miscarriage
- *Possible Contaminant:* Vegetables, milk, cheese, meat, seafood
- *Steps for Prevention:* Purchase pasteurized dairy products; cook foods properly; no cross-contamination; use sanitary practices.

Norwalk, Norwalk-like, or Norovirus

- *Cause of Illness:* Infection with Norwalk virus
- *Incubation Period:* Between 12 and 48 hours (average, 36 hours); duration, 12-60 hours
- *Symptoms:* Nausea, vomiting, diarrhea and abdominal cramps
- *Possible Contaminant:* Raw oysters/shellfish, water and ice, salads, frosting, person-to-person contact
- *Steps for Prevention:* Adequate and proper treatment and disposal of sewage, appropriate chlorination of water, restriction of infected food handlers from working with food until they no longer shed virus.

Salmonellosis

- *Cause of Illness:* Infection with Salmonella species
- *Incubation Period:* 12 to 24 hours
- *Symptoms:* Nausea, diarrhea, abdominal pain, fever, headache, chills, prostration
- *Possible Contaminant:* Meat, poultry, egg or milk products
- *Steps for Prevention:* Cook thoroughly; avoid cross-contamination; use sanitary practices.

Staphylococcus

- *Cause of Illness:* Toxin produced by certain strains of Staphylococcus aureus
- Incubation Period: One to six hours
- *Symptoms:* Severe vomiting, diarrhea, abdominal cramping
- *Possible Contaminant:* Custard- or cream-filled baked goods, ham, tongue, poultry, dressing, gravy, eggs, potato salad, cream sauces, sandwich fillings

- *Steps for Prevention:* Refrigerate foods; use sanitary practices.

Shigella
- *Cause of illness:* Water contaminated with human feces and unsanitary food handling
- *Incubation period:* 12 to 50 hours
- *Symptoms:* Abdominal pain, cramps, diarrhea, fever, vomiting, blood, and pus
- *Possible contaminants:* Salads, raw vegetables, dairy products, and poultry
- *Steps for prevention:* Practice proper washing and sanitizing techniques.

Toxoplasma Gondii
- *Cause of Illness:* Parasitic infection
- *Incubation Period:* Five to 23 days after exposure
- *Symptoms:* In healthy children and adults, toxoplasmosis may cause no symptoms at all, or may cause a mild illness (swollen lymph glands, fever, headache, and muscle aches).

Toxoplasmosis is a very severe infection for unborn babies and for people with immune system problems.

- *Possible Contaminant:* Cat, rodent or bird feces, raw or undercooked food.
- *Steps for Prevention:* Wash hands thoroughly after working with soil, cleaning litter boxes, before and after handling foods, and before eating. Cover sandboxes when not in use.

Vibrio
- *Cause of illness:* Excretion of toxin from infected fish and shellfish
- *Incubation period:* Four hours to four days
- *Symptoms:* Diarrhea, abdominal cramps, nausea, vomiting, headache, fever, and chills
- *Possible contaminants:* Fish and shellfish
- *Steps for prevention:* Cook fish and shellfish thoroughly

Yersiniosis
- *Cause of Illness:* Infection with Yersinia enterocolitica
- *Incubation Period:* One to three days

- *Symptoms:* Enterocolitis, may mimic acute appendicitis
- *Possible Contaminant:* Raw milk, chocolate milk, water, pork, other raw meats
- *Steps for Prevention:* Pasteurize milk; cook foods properly; no cross-contamination; use sanitary practices. Retail/Institutional Food Service Food Safety and Management

FOODBORNE DISEASE OUTBREAKS

An outbreak of foodborne illness occurs when a group of people consume the same contaminated food and two or more of the individuals develop the same symptoms or illness. For example, it may be a group that ate a meal together somewhere, or it may be a group of people who do not know each other at all, but who all happened to buy and eat the same contaminated item from a grocery store or restaurant.

For an outbreak to occur, an event or combination of events must happen to contaminate a batch of food eaten by a group of people. For example, contaminated food may be left out at room temperature for many hours, allowing the bacteria to multiply to high numbers, and then not properly cooked to kill the bacteria.

Many outbreaks are local in nature. For examples, a catered meal at a reception, a pot-luck supper, or eating a meal at an understaffed restaurant on a particularly busy day. These outbreaks are recognized when a group of people realise that they all became ill after a common meal, and someone calls the local health department. However, outbreaks are increasingly being recognized that are more widespread, that affect persons in many different places, and that are spread out over several weeks. As an illustration, in 2002, a salmonellosis outbreak was traced to persons who consumed ground beef in five states.

Forty seven cases were identified where 17 people were hospitalized and one died. The outbreak was recognized because it was caused by a multidrug-resistant *Salmonella* Newport and fingerprinting pattern of 94% of the isolates were indistinguishable indicating that the outbreak was originated by the same bacterial strain.

The vast majority of reported cases of foodborne illnesses is not part of recognized outbreaks, but occurs as individual or "sporadic" cases. It may be that many of these cases are actually part of unrecognized widespread or diffuse outbreaks. The initial clue that an outbreak is occurring can come in various ways:

- It may be when a person realises that several other people who were all together at an event have become ill and he or she calls the local health department.
- It may be when a physician realises she has seen more than the usual number of patients with the same illness.
- It may be when a county health department gets an unusually large number of reports of illness.

Once an outbreak is detected, an investigation begins. The outbreak is systematically described by time, place, and person by interviewing people, gathering epidemiological information, testing implicated food vehicle, and other associated information. If the causative microbe is not known, samples of stool or blood are collected from ill people and sent to the public health laboratory to make the diagnosis. Detecting and investigating such widespread outbreaks is a major challenge to our public health system. This is the reason that new and more sophisticated laboratory methods are being developed and used by CDC and in state public health department laboratories.

Epidemiology

One of the public health strategies for dealing with foodborne illness outbreaks is the use of epidemiology. Epidemiology is the study of factors determining and influencing the frequencies and distribution of a disease, injury, and other health-related events and their causes in a defined human population.

The purpose is to establish programmes to prevent and control their development and spread. Let's review a few very basic principles:
- The term "epidemic" is used when there is an occurrence of more cases of disease than expected in a given area or among a specific group of people over a particular period of time.
- The term "endemic" refers to the usual prevalence of a given disease or agent in a population or geographic area at all times. FSIS employs a group of epidemiologists to assist in investigating foodborne disease outbreaks related to meat, poultry, and egg products.

Surveillance Systems for Tracking Foodborne Diseases

The hardest outbreaks to detect are those that are spread over a large geographic area, with only a few cases in each state. These outbreaks can be detected by combining surveillance reports at the regional or national level and looking for increases in infections of a specific type. CDC is part of the US Public Health Service, with a mission to use the best scientific information to monitor, investigate, control and prevent public health problems.

CDC works closely with state health departments to monitor the frequency of specific diseases and conducts national surveillance for them. CDC provides expert epidemiologic and microbiologic consultation to health departments and other federal agencies on a variety of public health issues, including foodborne disease. CDC can also send a team into the field to conduct emergency field investigations of large or unusual outbreaks, in collaboration with state public health officials. CDC researchers develop new methods for identifying, characterizing and fingerprinting the microbes that cause disease. It translates laboratory research into practical field methods that can be used by public health authorities in states and counties.

CDC is not a regulatory agency. Government regulation of food safety is carried out by the Food and Drug Administration, the Food Safety and Inspection Service (USDA), the National Marine Fisheries Service, and other regulatory agencies.

CDC maintains regular contact with the regulatory agencies. Although it does not regulate the safety of food, the CDC assesses the effectiveness of current prevention efforts. It provides independent scientific assessment of what the problems are, how they can be controlled, and of where there are gaps in our knowledge.

FOOD PRESERVATION

Food preservation methods are intended to keep microorganisms out of foods, remove microorganisms from contaminated foods, and hinder the growth and activity of microorganisms already in foods. To keep microorganisms out of food, contamination is minimized during the entire food preparation process by sterilizing equipment, sanitizing it, and sealing products in wrapping materials. Microorganisms may be removed from liquid foods by filtering and sedimenting them or by washing and trimming them. Washing is particularly valuable for vegetables and fruits, and trimming is useful for meats and poultry products.

Heat

When heat is used to preserve foods, the number of microorganisms present, the microbial load, is an important consideration. Various types of microorganisms must also be considered because different levels of resistance exist. For example, bacterial spores are much more difficult to kill than vegetative bacilli. In addition, increasing acidity enhances the killing process in food preservation. Three basic heat treatments are used in food preservation: pasteurization, in which foods are treated at about 62°C for 30 minutes or 72°C for 15 to 17 seconds; hot filling, in which liquid foods and juices are boiled before being placed into containers; and steam treatment under pressure, such as used in the canning method. Each food preserved must be studied to determine how long it takes to kill the most resistant organisms present. The heat resistance of microorganisms is usually expressed as the thermal death time, the time necessary at a certain temperature to kill a stated number of particular microorganisms under specified conditions. In the canning process, the product is washed to remove soil. It is then blanched by a short period of exposure to hot water to deactivate enzymes in the food. Diseased parts in the food are removed, and the food is placed into cans by a filling machine. Sealed cans are then placed into a sterilizing machine called a retort, and the food is processed for a designated time and temperature.

Low temperatures

Low temperatures are used to preserve food by lowering microbial activity through the reduction of microbial enzymes. However, psychrophilic bacteria are known to grow even at cold refrigerator temperatures. These bacteria include members of the genera Pseudomonas, Alcaligenes, Micrococcus, and Flavobacterium. Fungi also grow at refrigeration temperatures. Slow freezing and quick freezing are used for long-term preservation. Freezing reduces the number of microorganisms in foods but does not kill them all. In microorganisms, cell proteins undergo denaturation due to increasing concentrations of solutes in the unfrozen water in foods, and damage is caused by ice crystals.

Chemicals

Several kinds of chemicals can be used for food preservation, including propionic acid, sorbic acid, benzoic acid, and sulfur dioxide. These acids are acceptable because they can be metabolized by the human body. Some antibiotics can also be used, depending upon local laws and ordinances. Tetracycline, for example, is often used to preserve meats. Storage and cooking normally eliminates the last remnants of antibiotic. In many foods,

Foodborne Pathogens

the natural acids act as preservatives. In sauerkraut, for example, lactic acid and acetic acid prevent contamination, while in fermented milks (yogurt, sour cream), acids perform the same function. For centuries, foods were prepared in this manner as a way of preventing microbial spoilage.

Drying

Drying is used to preserve food by placing foods in the sun and permitting the water to evaporate. Belt, tunnel, and cabinet dryers are used in industry for such things as instant coffee and cocoa. Freeze-drying, a process called lyophilization, is also valuable for producing a product free of moisture and very light.

Radiations

Ultraviolet radiation is valuable for reducing surface contamination on several foods. This short-wavelength light has been used in the cold storage units of meat processing plants. Ionizing radiations such as gamma rays can be used to preserve certain types of vegetables, fruits, and spices.

Microbial Determinations

To assess the presence and extent of microbial contamination in food, it is standard practice to perform several types of microbial determinations. These determinations are important because microorganisms from foods can cause such diseases as staphylococcal food poisoning, salmonellosis, typhoid fever, cholera, and gastroenteritis.

The standard plate count

One method for determining the number of bacteria in foods is the standard plate count. The procedure is performed by taking a gram of food sample and diluting it in 99-ml bottles of sterile water or buffer in the method described in the stage on aquatic microbiology. The number of bacterial colonies is multiplied by the dilution factor to determine the number of bacteria per gram of food.

Coliform and fungal determinations

In food, it is valuable to assess the number of coliform bacteria. These Gram-negative intestinal rods do not cause disease but are valuable indicators of fecal contamination. Presumably, when coliform bacteria are present, the food has been contaminated with fecal matter and is unfit to consume.

The coliform most commonly sought is Escherichia coli Various types of bacteriological media, such as violet red bile agar, are available for cultivating this bacteria, and the standard plate count technique can be performed to assess the number of coliform bacteria per gram of food. Fungi can be assessed in foods by using a medium such as Sabouraud dextrose agar. This medium encourages fungal spores to germinate and form visible masses of filaments called mycelia. A count of the resulting mycelia gives an estimate of the fungal contamination of the food.

The phosphatase test

To determine contamination in milk, the plate count technique can be used, but a more rapid test is the phosphatase test. Phosphatase is an enzyme destroyed by the pasteurization process. However, if the test for phosphatase shows that it is present after the milk has been treated, then the pasteurization has been unsuccessful. Testing for phosphatase is a more rapid and efficient method for determining contamination than the plate count technique.

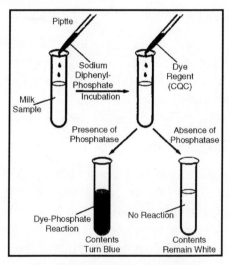

Fig. The Phosphatase Test.

The substrate sodium diphenyl phosphate is added to a pasteurized milk sample, and the tube is incubated. Then the dye reagent is added. If phosphatase is present, the tube contents turn blue. However, they remain clear if phosphatase was destroyed during the pasteurization process.

GM Food And Food Security

The official definition of food security, adopted at the World Food Summit of 1996, states:

- "Food security exists when all people, at all times, have physical and economic access to sufficient, safe and nutritious food to meet their dietary needs and food preferences for an active and healthy life."

This definition is understood within the framework of sustainability and was drawn from chapter 14.6 of Agenda 21, adopted at the 1992 United Nations Conference on Environment and Development, which states: "The major thrust of food security...is to bring about a significant increase in agricultural production in a sustainable way and to achieve a substantial improvement in people's entitlement to adequate food and culturally appropriate food supplies." The underlying assumption is that the means of increasing food availability in many countries exist, but are not being realised because of a range of constraints. In identifying and resolving these constraints, it is necessary to find sustainable ways of improving and reducing year-to-year variability in food production and open the way for broader food access.

The causes of food insecurity involve a complex interplay of economic, social, political and technical issues. An analysis of this interplay should determine the potential solution and best approach for a given population group. The issue for some communities is being able to produce sufficient food. For others, lack of money to purchase a wider selection of foods is the problem. Food insecurity and poverty are strongly correlated. The Swedish International Cooperation Agency defines poverty as a three-fold deficiency: a lack of security, ability and opportunity. Poverty is the main cause of food insecurity, and hunger is also a significant cause of poverty. Hunger is not only about quantity—it goes hand-in-hand with malnutrition. Food insecurity and malnutrition impair people's ability to develop skills and reduce their productivity.

A lag in farm productivity is closely associated with rural poverty and hunger. Food insecurity nevertheless is a reality experienced by the vulnerable in all societies and in all countries, developed and developing. In developed countries, the problem of food security is often a reflection of affordability and accessibility through conventional channels. Food security for the rural poor in developing countries is about producing or securing enough to feed one's household and being able to maintain that level of production year after year. Hunger and malnutrition increase susceptibility to disease and reduce people's ability to earn a livelihood. In instances where hunger is related to household income, improving food security by ensuring access to food or increasing the purchasing power of a family is essential. Providing poor communities with the skills to improve conditions in an economically and ecologically sustainable manner creates a window of opportunity to alleviate poverty at the subsistence-farming level and, on a larger scale, by having an impact on the economic development of the country.

A Potential Role for Modern Biotechnology

The Convention on Biological Diversity dictates the use and application of relevant technologies as a means of achieving the objectives of conservation and sustainable use with specific reference to biotechnology. Modern biotechnology is purported, from a technical perspective, to have a number of products for addressing certain food-security problems of developing countries. It offers the possibility of an agricultural system that is more reliant on biological processes rather than chemical applications.

The potential uses of modern biotechnology in agriculture include: increasing yields while reducing inputs of fertilizers, herbicides and insecticides; conferring drought or salt tolerance on crop plants; increasing shelf-life; reducing postharvest losses; increasing the nutrient content of produce; and delivering vaccines. The availability of such products could not only have an important role in reducing hunger and increasing food security, but also have the potential to address some of the health problems of the developing world. Achieving the improvements in crop yields expected in developing countries can help to alleviate poverty: directly by increasing the household incomes of small farmers who adopt these technologies; and indirectly, through spill-overs, as evidenced in the price slumps of herbicides and insecticides. Indirect benefits as a whole tend to have an impact on both technology adopters and non-adopters, the rural and urban poor. Indeed, some developing countries have identified priority areas such as tolerances to alkaline earth metals, drought and soil salinity, disease resistance, crop yields and nutritionally enhanced crops.

The adoption of technologies designed to prolong shelf-life could be valuable in helping to reduce postharvest losses in regionally important crops. Prime candidates in terms of crops of choice for development are the so-called 'orphan crops', such as cassava, sweet potato, millet, sorghum and yam. Multinationals have found no incentive to develop these crops and have instead invested in marketable crops with high profit returns. This strategy is intended to target wealthier farmers in temperate-zone countries with the financial capacity and tradition of supporting new seed products. However, here is a potential for multinational companies to develop crops grown largely in developing countries. The investment costs are low and the potential markets considerably large. While some public-sector research institutes in developing countries are forging ahead with the application of modern biotechnology, a small number are supported by government policy and therefore follow a defined agenda.

Still other governments believe that the risks associated with modern biotechnology outweigh the benefits. Currently, the many promises of modern biotechnology that could have an impact on food security have not been realised in most developing countries. In fact, the uptake of modern biotechnology has been remarkably low owing to the number of factors that underpin food security issues. In part, this could be because the first generation of commercially available crops using modern biotechnology were modified with single genes to impart agronomic properties with traits for pest and weed control, and not complex characteristics that would modify the growth of crops in harsh conditions. Secondly, the technologies are developed by companies in industrialized countries with little or no direct investment in, and which derive little economic benefit from, developing countries. Thirdly, many developing countries do not have the necessary biosafety frameworks to regulate the products of modern biotechnology.

For example, it took over two years for the Kenyan authorities to approve the field-testing of a virus-resistant sweet-potato variety because the scientific capacity for evaluating the product was unavailable. It should be noted, however, that such delays in the approval process have also been seen in developed countries, especially during the initiation of national regulatory evaluation. However, this trend is quickly changing as a number of developing countries either adopt or develop appropriate biotechnologies or regulatory infrastructures.

A report by the International Service for National Agricultural Research states that more than 40 crops are the focus of public-sector research programmes of 15 developing countries involving disease-resistant traits in rice, potato, maize, soybean, tomato, banana, papaya, sugarcane, alfalfa and plantain. For example, the Brazilian Agricultural Research Corporation has concentrated research into genetically modified crops on disease resistance in beans, papaya and potatoes.

The research programme at the University of Cape Town focuses on the development of crops resistant to viruses and to desiccation. The university has recently had a breakthrough with maize streak-virus resistance. In Thailand, the National Centre for Genetic Engineering and Biotechnology has supported research into disease resistance of rice, pepper and yard-long beans. therefore follow a defined agenda. Still other governments believe that the risks associated with modern biotechnology outweigh the benefits. Currently, the many promises of modern biotechnology that could have an impact on food security have not been realised in most developing countries. In fact, the uptake of modern biotechnology has been remarkably low owing to the number of factors that underpin food security issues.

Research Ownership

Research is a critical part of any effort aimed at improving food production and reducing poverty. Globally, much of agricultural R&D is carried out in the public sector, thereby serving the interests of developing countries. Public research in developed countries and Latin America is mostly conducted by government institutions and universities, whereas almost all agricultural research in Africa is carried out in public institutions, including R&D, technology transfer and dissemination of improved plant varieties. In general, international agricultural research institutions form a second level of research development and technology providers in developing countries. Public research institutes have, in the past, researched and improved orphan crops, mainly for donation to poor farmers or at cost. Generally, academic institutions are perceived as producers of knowledge that benefit and protect the public. Also, national and international research institutes aim to address the agricultural problems of resource-poor farmers in developing countries, *e.g.* increasing productivity through the use of a variety of techniques, including modern biotechnology. In reality, public institutions are now exposed to the forces of globalization and compelled to compete for their survival. In the current climate, government intervention in R&D worldwide has dwindled; hampering the level of innovation generated for public good. In fact, research facilities in many developing countries are poorly equipped, often limiting experiments to traditional and outdated research.

The diminishing role of public research institutes is perceived to have a major impact on the adoption of modern biotechnology in terms of introducing relevant products to those that need them most. Most of the field trials in the EU and the USA are conducted by private companies. An analysis of field-trial data from the USA shows that three crops account for 64% of all trials, of which 69% express herbicide- and pest-resistance traits. Of the trials conducted in the EU, 67% involve maize, sugar beet and rapeseed, and 71% of the novel gene categories presented herbicide- or pest-resistance traits. Less than 1% of all the trials in the EU and the USA are of plant varieties grown in tropical and subtropical climates, half of which have been conducted by the public sector.

Most of the public-sector research involving modern biotechnology in developing countries is still in the laboratory phase - none of the crops have progressed to marketable products. China, on the other hand, has approved the field-testing of over 500 GMOs to date and the commercial release of 50, including a prolonged shelf-life tomato, virus-resistant sweet pepper and vaccines for animal use. The experiences limiting the progression to commercializing research efforts range from: a lack of resources for meeting the high costs of regulatory requirements; lack of foresight, planning and business acumen for enabling the transition from research to a commercial product; and lack of capacity to negotiate patent licenses. Also, developments in modern biotechnology have occurred independently of the sustainable agricultural goals and priorities of the developing countries concerned. Furthermore, a needs assessment for a particular technology has often not been carried out before a research project is begun. Nevertheless, it is often argued that the commercialization of some products would encourage monocropping as national agricultural research has focused on a few crops, whereas rural communities tend to grow a wide range of crop species and plant varieties. The focus of international agricultural research centres is on plant production and protection, livestock production and health and food processing. With respect to food crops, research emphasis appears to be spread equally between cereals, root crops and legumes. However, within the cereals, research devoted to rice far outweighs research on maize and sorghum.

A large proportion of the total agricultural research activities of many developing countries are donor-funded. International research institutes, such as the Consultative Group on International Agricultural Research, depend on government grants and donations from philanthropic organizations for their survival, and yet investment in this sector has fallen in real terms. A significant fraction of the funding spent on international agricultural research institutes is used on activities covering a relatively large number of crops. The beneficiaries of such initiatives are a small number of countries with relatively advanced scientific capabilities. During the 1990s, developing countries as a group invested more in agricultural research than developed countries, even though the spending was unevenly distributed. In industrialized countries, private-sector investment in R&D far exceeds government spending on technology development, so that much of the public good previously entrusted to public research institutes is now privately owned. In comparison, private-sector investment in developing countries is around 1% of total global spending in this sector and developing countries invest less than 5% of the total private sector spending in biotechnology. Although the agricultural sector in developing countries is large and of significant importance to the domestic economy, spending in agricultural research does not match this level of activity. For instance, 80% of the food consumed in sub-Saharan Africa is obtained from domestic production.

IMPACT OF INTELLECTUAL PROPERTY RIGHTS ON RESEARCH

Intellectual property rights have been relevant to agriculture since their inception but have gained importance with regard to research in developed countries in the past 20-30 years. In particular, IPRs have been used to protect and preserve the value of products produced by conventional methods, *e.g.* the trademark registration of food products. The rationale for IPRs is that they encourage the inventor to advertise the invention and disclose the new knowledge, while simultaneously holding the rights to protect the invention from competitors. Disseminating this information is thought to stimulate new ideas and further rounds of innovation and technological advancement. IPRs afford time-limited protection to artistic, scientific, technological or economic products, and can be protected by way of copyrights, trademarks, design patents, utility patents, plant patents, plant breeders' rights and trade-secret law. Of these mechanisms, patents are considered the most powerful tool of the IPR system. Patents play different roles in different technologies and sectors. Patent protection of biotechnology makes it a tool for technology transfer and securing new markets in a global economy. Without protection, new ideas and information are entirely in the public domain.

This can, in certain systems, lead to underinvestment in R&D or the withholding of knowledge. Plant variety protection provides less protection than patents in that generally it makes provision for farmers' rights, allowing them to use harvested seed, and includes an exemption for research use. Despite the increase in availability, new plant varieties continue to be inaccessible or inappropriate for poor farmers, and the rate of innovation remains largely unchanged in countries with a PVP system. Studies have indeed shown that in middle-income countries, the principal beneficiaries of PVPs are commercial farmers and the seed industry. PVP is seen as a system that protects small advances in plant breeding, while a patent regime is thought to lead to the protection of big leaps in technological achievements. Patent protection for products of modern biotechnology is important because they are expensive to develop and easy to copy. Even so, developing countries have limited capabilities to innovate in industrial fields such as modern biotechnology, and to effectively enforce IPRs. A significant number of developing countries have not established intellectual property regimes that cover plants.

This situation may thus discourage private-sector investment. With no assurance that they can recoup some profit on GM products, multinationals are unlikely to devote much attention to the challenges of developing countries unless seen in a development aid context or through public-private partnership. Although this situation impedes private-sector investment in developing countries, it also implies that the freedom to operate on products designated for local markets is not hindered. Exercising this freedom to operate is not a well-understood concept. For example, in the case of 'golden rice', where permission was required for the use of about 70 patents, the impression was that the patents were being relinquished in favour of the poor. In fact, most of the patents involved are not valid in the major rice-consuming countries.

The technology donated for the development of virus-resistance in non-commercial potato varieties is free of patents relevant to Mexico, and the same holds true in the case of virus-resistant sweet potatoes in Kenya. Researchers are usually unaware of the proprietary status of the technologies they are using in their work. Nevertheless, the proliferation of broad patents is thought to impede the research capabilities of other interested parties. Some countries grant very broad patents conferring monopoly rights over large areas of research, thereby potentially threatening the other goal of intellectual property, namely the right to build upon the original invention. The prevailing patent rules have the potential to limit the accessibility of these technologies to public institutions and ultimately poor farmers.

Furthermore, the strengthening of IPRs is thought to restrict the flow of germplasm and inhibit the development of new plant varieties. This is because if and when researchers in public institutions do get permission to develop the technologies further, access is granted under licence agreements with restrictions on commercializing innovations. It is also argued that a stringent, multilateral IPR system will not benefit all countries equally. Indeed, the benefits will largely be influenced by the economic and technological levels of development in each country.

The United Kingdom Commission on Intellectual Property Rights, "the critical issue in respect of IPRs is perhaps not whether it promotes trade or foreign investment, but how it helps or hinders developing countries to gain access to technologies that are required for their development." During the 1990s, obtaining a patent application in the USA cost US$20,000 and twice as much in the EU. In general, PVP is cheaper; valued at one-tenth of the patent price. Moreover, the preparation of a food-safety dossier for a product derived from modern biotechnology, for example, is estimated to be around US$1 million.

These estimates cannot be compared with the regulatory costs in developing countries. The cost of regulation in developing countries does not encourage the commercialization of products of modern biotechnology developed by public-sector research institutes. In most cases, the regulatory costs far exceed the research costs. Many developing countries do not have the resources to match private-sector investment in modern biotechnology. In this new playing field, public institutions also need resources to deal with intellectual property rights to help compensate and increase the public benefit. Otherwise their involvement in R&D could be deterred by lack of funding. If public institutions are to use the techniques of modern biotechnology, then the use of IPRs as a framework for facilitating technology transfer must be emphasized more than its handling as a revenue-generating system. IPRs can, however, play a major role in clarifying the mechanisms for access to technology, and determine the downstream aspects of use and exploitation of genetic resources.

There are several ways in which public institutions and small companies in developing countries can gain access to patented genes and enabling technologies to overcome the current barriers to research. The first of these includes a measure of goodwill by multinationals to relinquish their rights to technologies for use by researchers in developing countries by adopting programmes of social responsibility as in the case of 'golden rice', a variety containing beta-carotene, virus-resistant sweet potatoes and a virus-resistant non-commercial potato variety in Mexico.

A different type of programme initiated in the USA has established an intellectual property clearing house to make information on the intellectual property owned by public research institutes, including universities, available to researchers around the world. There are also suggestions that redesigning patent laws to narrow the type and scope of patent coverage ought to make more technologies accessible to public institutions. The thinking behind these suggestions is that applying a stronger standard for rejecting patent applications for inventions that are 'obvious' should deter the patenting of minor inventions. In addition, a law that requires an invention to be genuinely useful in theory should reduce the number of patent applications being submitted. At present, it is possible in some countries to submit patent applications for abstract concepts that potentially protect large areas of research and thereby exclude innovation by others. Another option that may be attractive to developing countries is the creation of collaborations that involve research institutes, universities and the private sector.

The nature of these collaborations is likely to be influenced by the level of expertise and resources within the national public research institutes. Where a solid knowledge base exists, the public partners may be in a position to develop or acquire a technology that could be transferred into locally adapted varieties. Smaller institutes are more likely to provide the genetic resources and a positive public image. It is believed that such alliances would benefit public institutions and private companies; offering them an opportunity to license and distribute the technology.

The most well-known public-private sector partnerships include organizations such as the International Service for the Acquisition of Agri-biotech Applications that negotiate access to private-sector technologies for the improvement of subsistence crops and/or the transfer of technology and know-how. Although several types of public-private sector alliances already exist, the two newly established initiatives worth mentioning are the Global Cassava Partnership and the African Agricultural Technology Foundation, launched in November 2002 by FAO and in March 2003 by the Rockefeller Foundation, respectively. The former is a partnership involving some of the world's leading experts in cassava research, working mainly in public institutions. The AATF intends to function as a clearing house of available technologies with the primary aim of improving food security and reducing the poverty situation of small farmers, by facilitating transfer and use of appropriate technologies.

Such licensing arrangements have been put to the test in other fields. However, as in the case of the AATF, a clearing house is required to acquire the necessary technologies and permit their further use for developing-country needs. The drawback may be a requirement to divide the commercial sector into subsistence, middle-income and commercial markets. This market division may not be easy to achieve as some large developing countries have both commercially important markets and subsistence farmers.

A briefing document commissioned by the United Nations Industrial Development Organization proposes six activities that build upon private-sector investment and enable biotechnology transfer:

- Enabling government policies;
- Access to up-to-date authoritative information;
- Regional brokering service to strengthen public-private partnerships;
- A regional biotechnology investment service;
- An international intellectual property escrow service; and
- Initiatives for risk-shifting.

Each of the proposals can be implemented as a stand-alone project or in combination, as best suits the national and/or regional situation.

10

Defining Food Security

Food security is a flexible concept as reflected in the many attempts at definition in research and policy usage. Even a decade ago, there were about 200 definitions in published writings. Whenever the concept is introduced in the title of a study or its objectives, it is necessary to look closely to establish the explicit or implied definition. The continuing evolution of food security as an operational concept in public policy has reflected the wider recognition of the complexities of the technical and policy issues involved.

The most recent careful redefinition of food security is that negotiated in the process of international consultation leading to the World Food Summit in November 1996. The contrasting definitions of food security adopted in 1974 and 1996, along with those in official FAO and World Bank documents of the mid-1980s, are set out below with each substantive change in definition underlined. A comparison of these definitions highlights the considerable reconstruction of official thinking on food security that has occurred over 25 years. These statements also provide signposts to the policy analyses, which have reshaped our understanding of food security as a problem of international and national responsibility.

Food security as a concept originated only in the mid-1970s, in the discussions of international food problems at a time of global food crisis. The initial focus of attention was primarily on food supply problems - of assuring the availability and to some degree the price stability of basic foodstuffs at the international and national level. That supply-side, international and institutional set of concerns reflected the changing organization of the global food economy that had precipitated the crisis. A process of international negotiation followed, leading to the World Food Conference of 1974, and a new set of institutional arrangements covering information, resources for promoting food security and forums for dialogue on policy issues. The issues of famine, hunger and food crisis were also being extensively examined, following the events of the mid-1970s. The outcome was a redefinition of food security, which recognised that the behaviour of potentially vulnerable and affected people was a critical aspect.

A third, perhaps crucially important, factor in modifying views of food security was the evidence that the technical successes of the Green Revolution did not automatically and rapidly lead to dramatic reductions in poverty and levels of malnutrition. These problems were recognised as the result of lack of effective demand.

Official concepts of food security

The initial focus, reflecting the global concerns of 1974, was on the volume and stability of food supplies.

Food security was defined in the 1974 World Food Summit as:
- "Availability at all times of adequate world food supplies of basic foodstuffs to sustain a steady expansion of food consumption and to offset fluctuations in production and prices".

In 1983, FAO expanded its concept to include securing access by vulnerable people to available supplies, implying that attention should be balanced between the demand and supply side of the food security equation:
- "Ensuring that all people at all times have both physical and economic access to the basic food that they need".

In 1986, the highly influential World Bank report "Poverty and Hunger" focused on the temporal dynamics of food insecurity. It introduced the widely accepted distinction between chronic food insecurity, associated with problems of continuing or structural poverty and low incomes, and transitory food insecurity, which involved periods of intensified pressure caused by natural disasters, economic collapse or conflict.

This concept of food security is further elaborated in terms of:
- "Access of all people at all times to enough food for an active, healthy life".

By the mid-1990s food security was recognised as a significant concern, spanning a spectrum from the individual to the global level. However, access now involved sufficient food, indicating continuing concern with protein-energy malnutrition. But the definition was broadened to incorporate food safety and also nutritional balance,

Defining Food Security

reflecting concerns about food composition and minor nutrient requirements for an active and healthy life.

Food preferences, socially or culturally determined, now became a consideration. The potentially high degree of context specificity implies that the concept had both lost its simplicity and was not itself a goal, but an intermediating set of actions that contribute to an active and healthy life.

The 1994 UNDP Human Development Report promoted the construct of human security, including a number of component aspects, of which food security was only one. This concept is closely related to the human rights perspective on development that has, in turn, influenced discussions about food security.

The 1996 World Food Summit adopted a still more complex definition:
- "Food security, at the individual, household, national, regional and global levels [is achieved] when all people, at all times, have physical and economic access to sufficient, safe and nutritious food to meet their dietary needs and food preferences for an active and healthy life".

This definition is again refined in The State of Food Insecurity 2001:
- "Food security a situation that exists when all people, at all times, have physical, social and economic access to sufficient, safe and nutritious food that meets their dietary needs and food preferences for an active and healthy life".

This new emphasis on consumption, the demand side and the issues of access by vulnerable people to food, is most closely identified with the seminal study by Amartya Sen. Eschewing the use of the concept of food security, he focuses on the entitlements of individuals and households. The international community has accepted these increasingly broad statements of common goals and implied responsibilities.

But its practical response has been to focus on narrower, simpler objectives around which to organise international and national public action. The declared primary objective in international development policy discourse is increasingly the reduction and elimination of poverty.

The 1996 WFS exemplified this direction of policy by making the primary objective of international action on food security halving of the number of hungry or undernourished people by 2015. Essentially, food security can be described as a phenomenon relating to individuals. It is the nutritional status of the individual household member that is the ultimate focus, and the risk of that adequate status not being achieved or becoming undermined. The latter risk describes the vulnerability of individuals in this context.

Vulnerability may occur both as a chronic and transitory phenomenon. Food security exists when all people, at all times, have physical, social and economic access to sufficient, safe and nutritious food which meets their dietary needs and food preferences for an active and healthy life. Household food security is the application of this concept to the family level, with individuals within households as the focus of concern. Food insecurity exists when people do not have adequate physical, social or economic access to food.

THE CHALLENGES TO FOOD SECURITY

In developing countries, 800 million people are undernourished, of which a significant proportion live on less than US$1 per day, despite a more than 50% decline in world food prices over the past 20 years. Global food production has soared, making a variety of foods available to all consumers. Although the decline in food prices in developed countries has benefited the poor who spend a considerable share of their income on food, this trend has not had much impact on the majority in the developing world, with sub-Saharan Africa painting the gloomiest picture.

Due to the substantial price reduction in this commodity sector, cereals have become the staple foods in the diet of poor people. While yield increases in the major cereal crops has meant more calorific intake of food, micronutrient malnutrition remains a serious problem. Regional analyses depict sub-Saharan Africa as the only region where both the number and proportion of malnourished children have consistently risen in the past three decades. However, malnutrition in South Asia is also very high.

The world population is projected to reach 8 billion by 2025, and it is estimated that most of this growth will occur in developing countries. Feeding and housing an additional 2 billion people will cause considerable pressure on land, water, energy and other natural resources. Looking at projections to 2020, the worldwide per capita availability of food is expected to increase by approximately 7%, *i.e.* 2900 calories per person per day. Nonetheless, an average availability of 2300 calories is projected for individuals in sub-Saharan Africa, a figure that is just above the recommended minimum calorie intake for an active and productive life.

Defining Food Security

In terms of agricultural output, preliminary world estimates for 2001 suggested that growth was as low as 0.6%. Annual rates also demonstrate a trend of decreased productivity, particularly in developing-country regions. Output growth in Asia has systematically declined over the past five years the rates experienced in sub-Saharan Africa are lower than average. Agricultural productivity is important for food security in that it has an impact on food supplies, prices, and the incomes and purchasing power of farmers. Improving food security at the national level requires an increase in the availability of food through increased agricultural production, or by increasing imports.

To augment domestic production and maintain an adequate supply of food, food-insecure countries often rely on imports and food aid. Export earnings are frequently low and do not suffice to provide foreign exchange to finance imports. Thus, in the long term, importing food is unsustainable. Historically, increased food production in the developing countries can be attributed to the cultivation of more land rather than to the deployment of improved farming practices or to the application of new technologies. By its very nature, agriculture threatens other ecosystems, a situation that can be exacerbated by over-cultivation, overgrazing, deforestation and bad irrigation practices. However, increased demands for food in Asia, Europe and North Africa have to be met by increasing yields because most land in these areas is already used for agriculture.

The potential to expand agricultural land exists in Latin America and sub-Saharan Africa only, where much of the remaining land is marginal for agricultural expansion. The implication, therefore, is that the increase in food production needed to feed the world's growing population can only be met by increasing the amount of food produced per hectare. Recognizing the extent of environmental degradation caused mainly by human activities, the multilateral agreements that arose from the UNCED meeting of 1992 were intended to address the compromised food-security situation on a global scale. One such agreement is the United Nations Convention to Combat Desertification. This agreement promotes the implementation of practices intended to reverse desertification for sustainable land use and food security. As the more affluent developed countries tend to produce more food, some argue that redistribution of these surpluses could feed the escalating populations of developing countries. Redistribution, however, requires policy changes that may be impossible to implement on a global scale. Therefore, a substantial proportion of the food demands of developing countries will have to be met by the agricultural systems in these countries. Enabling a consistent and sustainable supply of food will require an overhaul in the production processes and the supporting infrastructure. Finding solutions to declining crop yields requires an effort that will improve the assets on which agriculture relies; namely, soils, water and biodiversity.

Transforming the agricultural systems of rural farmers by introducing technologies that integrate agro-ecological processes in food production, while minimizing adverse effects to the environment, is key to sustainable agriculture. In addition, increases in crop yield must be met with the use of locally available low-cost technologies and minimum inputs without causing damage to the environment.

Attaining Food Security

Within the context of the definition, three distinct components appear central to attaining food security: availability, accessibility and adequacy. Within each component, questions are raised which may need to be addressed to improve the food security situation at a national, regional or international level.

The questions raised here are intended to demonstrate the complexity of the issue and are by no means exhaustive:

- *Availability*: is there enough food available through domestic production or imports to meet the immediate needs? Is production environmentally sustainable to meet long-term demands? Are the distribution systems effective in reaching low-income and rural communities?
- *Accessibility*: do the vulnerable in society have the purchasing power to attain food security? Can they afford the minimum basic diet of 2100 calories per day required for an active and productive life?
- Adequacy: does the food supply provide for the differing nutritional needs, *i.e.* a balanced diet, offering the necessary variety of foods at all times? Is the food properly processed, stored and prepared?

Global food productivity is undergoing a process of rapid transformation as a result of technological progress in the fields of communication, information, transport and modern biotechnology. A general observation is that technologies tend to be developed in response to market pressures, and not to the needs of the poor who have no purchasing power. As agriculture is the main economic activity of rural communities, optimizing the levels of production will generate employment and income, and thus uplift the wealth and well-being of the community.

Improving agricultural production in developing countries is fundamental to reducing poverty and increasing food security. Investment to raise agricultural productivity can be achieved through the introduction of superior technologies such as better-quality seeds, crop rotation systems etc. It is argued, however, that the adoption of earlier agricultural technologies has led to the emergence of more virulent strains of pests, pathogens and weeds, soil deterioration and a loss of biodiversity.

The Green Revolution, in particular, focused on wheat and rice—not much attention was paid to staple crops such as sorghum, cassava or millet. Also, the seeds and fertilizers required to grow the higher-yielding varieties were expensive and therefore not accessible to all. Reaffirming support for the principles agreed upon at the UNCED, the United Nations Millennium Development Goals have set a road map for protecting the environment. Embraced in these time-bound goals is a new development ethic that demands sustainability in a framework where progress is measured in terms of actions reconciling the economic and ecological factors of food production for the benefit of present and future generations?

Extending this understanding to agriculture, sustainable agriculture is defined as:
- Environmentally sound, preserving resources and maintaining production potential;
- Profitable for farmers and workable on a long-term basis;
- Providing food quality and sufficiency for all people;
- Socially acceptable; and
- Socially equitable, between different countries and within each country.

A secure food system is one in which the ecological resources on which food production depends allow for their continued use, with minimum damage for present and future generations.

In other words, food security and sustainable agriculture are interlinked and both are central to the concept of sustainable development. The FAO Anti-hunger Programme reported that increasing investment in agriculture and rural development can reduce hunger. To reduce the number of hungry people by half by 2015, it estimated that funding of US$24 billion would be required for agricultural research, emergency food assistance, and improving rural infrastructure.

By contrast, at the current rate of progress, the number of food-insecure would fall by only 24%. For people to be food-secure, they must have access to the resources needed to buy or produce their own food.

Breaking the poverty cycle of rural communities, whose livelihoods depend on agriculture, will require investment in different technologies to address the different constraints experienced in different world regions. The production problems experienced by farmers vary between countries and communities, and technological solutions need to be relevant to those circumstances, *i.e.* one solution will not be suitable everywhere.

The potential of some of these technologies has been demonstrated in various world regions— for instance, agro-ecological improvement programmes involving:

- Better harvesting and conservation of water, even in rain-fed environments;
- A reduction of soil erosion by adopting zero tillage combined with the use of green manure and herbicides as in Argentina and Brazil; and
- Pest and weed control without pesticide or herbicide use, *e.g.* Bangladesh and Kenya, has been well tested and established.

Indeed, such programmes are now widely accepted as being at the core of sustainable agriculture. The communities that participated in these projects were able to transform food production through the use of resource-management strategies that focused on improving the soil by growing leguminous crops and applying agro-forestry, zero tillage and green manure. These and other projects have proved that the sustainability of any farming practice and the conditions under which production can be maintained at reasonable levels cannot be predicted with absolute certainty. Some regions may be better able to transfer high-yielding technologies with varying degrees of success.

The uptake of new production systems has proved successful where the programmes have included the participation of entire communities and have not been introduced to isolated groups of farmers. Producing nutritionally enhanced properties in staple crops eaten by the poor could reduce the burden of disease in many developing countries. Scientists at the International Crops Research Institute for the Semi-Arid Tropics have developed a pearl millet variety enhanced with beta-carotene. The trait naturally occurs in two Burkina Faso millet lines from which it was transferred by conventional breeding methods. Genetic modification of japonica rice with a ferritin gene has not given superior results compared

Defining Food Security

to rice with an 80% increase in iron density produced by conventional plant breeding at the International Rice Research Institute. Research and technology alone will not drive agricultural growth. Inadequate infrastructure and poorly functioning markets tend to exacerbate the problem of food insecurity. The cost of marketing farm produce can be prohibitive for small-scale farmers, as their isolation prevents the link between agricultural and non-agricultural activities among adjacent villages and between rural and urban areas.

Building roads in rural areas is vital to facilitating growth, trade and exchange of farm and non-farm products in rural communities, even those that can adequately feed themselves. For instance, government investment in irrigation projects, storage and transport facilities, roads connecting villages to larger markets in the rural areas of China and India, has made an impressive impact on employment and productivity and ultimately provided opportunities for poverty alleviation in the affected areas. UNDP, basic thresholds in roads, power, ports and communications must be reached in order to sustain growth.

CLIMATE CHANGE AND FOOD SECURITY
Impacts on Food Production and Availability

Climate change affects agriculture and food production in complex ways. It affects food production directly through changes in agro-ecological conditions and indirectly by affecting growth and distribution of incomes, and thus demand for agricultural produce. Impacts have been quantified in numerous studies and under various sets of assumptions. A selection of these results is presented in *Quantifying the Impacts on Food Security*. Here it is useful to summarize the main alterations in the agro-ecological environment that are associated with climate change.

Changes in temperature and precipitation associated with continued emissions of greenhouse gases will bring changes in land suitability and crop yields. In particular, the Intergovernmental Panel on Climate Change (IPCC) considers four families of socio-economic development and associated emission scenarios, known as Special Report on Emissions Scenarios (SRES) A2, B2, A1, and B1. Of relevance to this review, of the SRES scenarios, A1, the "business-as-usual scenario," corresponds to the highest emissions, and B1 corresponds to the lowest.

The other scenarios are intermediate between these two. Importantly for agriculture and world food supply, SRES A2 assumes the highest projected population growth of the four (United Nations high projection) and is thus associated to the highest food demand. Depending on the SRES emission scenario and climate models considered, global mean surface temperature is projected to rise in a range from 1.8°C (with a range from 1.1°C to 2.9°C for SRES B1) to 4.0°C (with a range from 2.4°C to 6.4°C for A1) by 2100.

In temperate latitudes, higher temperatures are expected to bring predominantly benefits to agriculture: the areas potentially suitable for cropping will expand, the length of the growing period will increase, and crop yields may rise.

A moderate incremental warming in some humid and temperate grasslands may increase pasture productivity and reduce the need for housing and for compound feed. These gains have to be set against an increased frequency of extreme events, for instance, heat waves and droughts in the Mediterranean region or increased heavy precipitation events and flooding in temperate regions, including the possibility of increased coastal storms; they also have to be set against the fact that semiarid and arid pastures are likely to see reduced livestock productivity and increased livestock mortality.

In drier areas, climate models predict increased evapotranspiration and lower soil moisture levels. As a result, some cultivated areas may become unsuitable for cropping and some tropical grassland may become increasingly arid. Temperature rise will also expand the range of many agricultural pests and increase the ability of pest populations to survive the winter and attack spring crops.

Another important change for agriculture is the increase in atmospheric carbon dioxide (CO_2) concentrations. Depending on the SRES emission scenario, the atmospheric CO_2 concentration is projected to increase from H"379 ppm today to >550 ppm by 2100 in SRES B1 to >800 ppm in SRES A1FI. Higher CO_2 concentrations will have a positive effect on many crops, enhancing biomass accumulation and final yield. However, the magnitude of this effect is less clear, with important differences depending on management type (*e.g.*, irrigation and fertilization regimes) and crop type. Experimental yield response to elevated CO_2 show that under optimal growth conditions, crop yields increase at 550 ppm CO_2 in the range of 10 per cent to 20 per cent for C_3 crops (such as wheat, rice, and soybean), and only 0–10 per cent for C_4 crops such as maize and sorghum. Yet the nutritional quality of agricultural produce may not increase in line with higher yields. Some cereal and forage crops, for example, show lower protein concentrations under elevated CO_2 conditions.

Finally, a number of recent studies have estimated the likely changes in land suitability, potential yields, and agricultural production on the current suite of crops and cultivars available today. Therefore, these estimates implicitly include adaptation using available management techniques and crops, but excluding new cultivars from breeding or biotechnology. These studies are in essence based on the FAO/ International Institute for Applied Systems Analysis (IIASA) agro-ecological zone (AEZ) methodology. For instance, pioneering work in ref. 9 suggested that total land and total prime land would remain virtually unchanged at the current levels of 2,600 million and 2,000 million hectares (ha), respectively. The same study also showed, however, more pronounced regional shifts, with a considerable increase in suitable cropland at higher latitudes (developed countries +160 million ha) and a corresponding decline of potential cropland at lower latitudes (developing countries "110 million ha). An even more pronounced shift within the quality of cropland is predicted in developing countries. The net decline of 110 million ha is the result of a massive decline in agricultural prime land of H"135 million ha, which is offset by an increase in moderately suitable land of >20 million ha. This quality shift is also reflected in the shift in land suitable for multiple cropping. In sub-Saharan Africa alone, land for double cropping would decline by between 10 million and 20 million ha, and land suitable for triple copping would decline by 5 million to 10 million ha. At a regional level, similar approaches indicate that under climate change, the biggest losses in suitable cropland are likely to be in Africa, whereas the largest expansion of suitable cropland is in the Russian Federation and Central Asia.

Impacts on the Stability of Food Supplies

Global and regional weather conditions are also expected to become more variable than at present, with increases in the frequency and severity of extreme events such as cyclones, floods, hailstorms, and droughts. By bringing greater fluctuations in crop yields and local food supplies and higher risks of landslides and erosion damage, they can adversely affect the stability of food supplies and thus food security.

Neither climate change nor short-term climate variability and associated adaptation are new phenomena in agriculture, of course. As shown, for instance, in ref. 9, some important agricultural areas of the world like the Midwest of the United States, the northeast of Argentina, southern Africa, or southeast Australia have traditionally experienced higher climate variability than other regions such as central Africa or Europe. They also show that the extent of short-term fluctuations has changed over longer periods of time. In developed countries, for instance, short-term climate variability increased from 1931 to 1960 as compared with 1901 to 1930, but decreased strongly in the period from 1961 to 1990. What is new, however, is the fact that the areas subject to high climate variability are likely to expand, whereas the extent of short-term climate variability is likely to increase across all regions. Furthermore, the rates and levels of projected warming may exceed in some regions the historical experience.

If climate fluctuations become more pronounced and more widespread, droughts and floods, the dominant causes of short-term fluctuations in food production in semiarid and subhumid areas, will become more severe and more frequent. In semiarid areas, droughts can dramatically reduce crop yields and livestock numbers and productivity. Again, most of this land is in sub-Saharan Africa and parts of South Asia, meaning that the poorest regions with the highest level of chronic undernourishment will also be exposed to the highest degree of instability in food production. How strongly these impacts will be felt will crucially depend on whether such fluctuations can be countered by investments in irrigation, better storage facilities, or higher food imports. In addition, a policy environment that fosters freer trade and promotes investments in transportation, communications, and irrigation infrastructure can help address these challenges early on.

Impacts of Climate Change on Food Utilization

Climate change will also affect the ability of individuals to use food effectively by altering the conditions for food safety and changing the disease pressure from vector, water, and food-borne diseases. The IPPC Working Group II provides a detailed account of the health impacts of climate change of its fourth assessment report. It examines how the various forms of diseases, including vector-borne diseases such as malaria, are likely to spread or recede with climate change. This chapter focuses on a narrow selection of diseases that affect food safety directly, *i.e.*, food and water-borne diseases.

The main concern about climate change and food security is that changing climatic conditions can initiate a vicious circle where infectious

Defining Food Security

disease causes or compounds hunger, which, in turn, makes the affected populations more susceptible to infectious disease. The result can be a substantial decline in labour productivity and an increase in poverty and even mortality. Essentially all manifestations of climate change, be they drought, higher temperatures, or heavy rainfalls have an impact on the disease pressure, and there is growing evidence that these changes affect food safety and food security.

The recent IPCC report also emphasizes that increases in daily temperatures will increase the frequency of food poisoning, particularly in temperate regions. Warmer seas may contribute to increased cases of human shellfish and reef-fish poisoning (ciguatera) in tropical regions and a poleward expansion of the disease. However, there is little new evidence that climate change significantly alters the prevalence of these diseases. Several studies have confirmed and quantified the effects of temperature on common forms of food poisoning, such as salmonellosis. These studies show an approximately linear increase in reported cases for each degree increase in weekly temperature. Moreover, there is evidence that temperature variability affects the incidence of diarrhoeal disease. A number of studies found that rising temperatures were strongly associated with increased episodes of diarrhoeal disease in adults and children. These findings have been corroborated by analyses based on monthly temperature observations. Several studies report a strong correlation between monthly temperature and diarrhoeal episodes on the Pacific Islands, Australia, and Israel.

Extreme rainfall events can increase the risk of outbreaks of water-borne diseases particularly where traditional water management systems are insufficient to handle the new extremes. Likewise, the impacts of flooding will be felt most strongly in environmentally degraded areas, and where basic public infrastructure, including sanitation and hygiene, is lacking. This will raise the number of people exposed to water-borne diseases (*e.g.*, cholera) and thus lower their capacity to effectively use food.

Impacts of Climate Change on Access to Food

Access to food refers to the ability of individuals, communities, and countries to purchase sufficient quantities and qualities of food. Over the last 30 years, falling real prices for food and rising real incomes have led to substantial improvements in access to food in many developing countries. Increased purchasing power has allowed a growing number of people to purchase not only more food but also more nutritious food with more protein, micronutrients, and vitamins. East Asia and to a lesser extent the Near-East/North African region have particularly benefited from a combination of lower real food prices and robust income growth. From 1970 to 2001, the prevalence of hunger in these regions, as measured by FAO's indicator of undernourishment, has declined from 24 per cent to 10.1 per cent and 44 per cent to 10.2 per cent respectively. In East Asia, it was endogenous income growth that provided the basis for the boost in demand for food, which was largely produced in the region; in the Near-East North African region demand was spurred by exogenous revenues from oil and gas exports, and additional food supply came largely from imports. But in both regions, improvements in access to food have been crucial in reducing hunger and malnutrition.

FAO's longer-term outlook to 2050 suggests that the importance of improved demand side conditions will even become more important over the next 50 years. The regions that will see the strongest reductions in the prevalence of undernourishment are those that are expected to see the highest rates of income growth. South Asia stands to benefit the most. Spurred by high income growth the region is expected to reduce the prevalence of undernourishment from >22 per cent currently to 12 per cent by 2015 and just 4 per cent by 2050. Progress is also expected for sub-Saharan Africa, but improvements will be less pronounced and are expected to set in later.

Over the next 15 years, for instance, the prevalence of undernourishment will decline less than in other regions, from H-33 per cent to a still worrisome 21 per cent, as significant constraints (soil nutrients, water, infrastructure, etc.) will limit the ability to further increase food production locally, while continuing low levels of income rule out the option of importing food. In the long run, however, sub-Saharan Africa is expected to see a more substantial decline in hunger; by 2050, <6 per cent of its total population are expected to suffer from chronic hunger. It is important to note that these FAO projections do not take into account the effects of climate change.

By coupling agro-ecologic and economic models, others have gauged the impact of climate change on agricultural gross domestic product (GDP) and prices. At the global level, the impacts of climate change are likely to be very small; under a range of SRES and associated climate-change scenarios, the estimates range from a decline of "1.5 per cent to an increase of +2.6 per cent by 2080.

Defining Food Security

At the regional level, the importance of agriculture as a source of income can be much more important. In these regions, the economic output from agriculture itself will be an important contributor to food security. The strongest impact of climate change on the economic output of agriculture is expected for sub-Saharan Africa, which means that the poorest and already most food-insecure region is also expected to suffer the largest contraction of agricultural incomes. For the region, the losses in agricultural GDP, compared with no climate change, range from 2 per cent to 8 per cent for the Hadley Centre Coupled Model, version 3 and 7–9 per cent for the Commonwealth Scientific and Industrial Research Organisation projections.

Impacts on Food Prices

Essentially all SRES development paths describe a world of robust economic growth and rapidly shrinking importance of agriculture in the long run and thus a continuation of a trend that has been underway for decades in many developing regions. SRES scenarios describe a world where income growth will allow the largest part of the world's population to address possible local production shortfalls through imports and, at the same time, find ways to cope with safety and stability issues of food supplies. It is also a world where real incomes rise more rapidly than real food prices, which suggests that the share of income spent on food should decline and that even high food prices are unlikely to create a major dent in the food expenditures of the poor. However, not all parts of the world perform equally well in the various development paths and not all development paths are equally benign for growth. Where income levels are low and shares of food expenditures are high, higher prices for food may still create or exacerbate a possible food security problem.

There are a number of studies that have ventured to measure the likely impacts of climate change on food prices. The basic messages that emerge from these studies are: first, on average, prices for food are expected to rise moderately in line with moderate increases of temperature (until 2050); some studies even foresee a mild decline in real prices until 2050. Second, after 2050 and with further increases in temperatures, prices are expected to increase more substantially. In some studies and for some commodities (rice and sugar) prices are forecast to increase by as much as 80 per cent above their reference levels without climate change. Third, price changes expected from the effects of global warming are, on average, much smaller than price changes from socio-economic development paths. For instance, the SRES A2 scenario would imply a price increase in real cereal prices by H-170 per cent. The (additional) price increase caused by climate change (in the Hadley Centre Coupled Model, version 3, climate change case) would only be 14.4 per cent. Overall, this appears to be the sharpest price increase reported and it is not surprising that this scenario would imply a stubbornly high number of undernourished people until 2080. However, it is also needless to say that a constant absolute number of undernourished people would still imply a sharp decline in the prevalence of hunger, and, given the high population assumptions in the SRES A2 world (>13.6 billion people globally and >11.6 billion in the developing world) this would imply a particularly sharp drop in the prevalence from currently 17 per cent to H-7 per cent by 2080.

QUANTIFYING THE IMPACTS ON FOOD SECURITY

A number of studies have recently quantified the impacts of climate change on food security. In terms of quantifying agronomic yield change projections, these studies are either based on the AEZ tools developed by the IIASA analysis or the Decision Support System for Agrotechnology Transfer suite of crop models; all use the IIASA-Basic Linked System (BLS) economic model for assessing economic impacts. These tools, with some modifications relating to how crop yield changes are simulated, have also been used by others to undertake similar assessments and provide sensitivity analyses across a range of SRES and general circulation model (GCM) projections. Many other simulations have also examined the effects of climate change with and without adaptation (induced technological progress, domestic policy change, international trade liberalization, etc.), with and without mitigation (*e.g.*, CO_2 stabilization, variants for temperature, rainfall change and distribution) or provide impact assessments for different speeds of climate change. The quantitative results for food security, trying to illuminate some of the differences and extract the main messages that emerge from the various studies. Unless indicated, all simulation results discussed below include the combined effects of climate change and elevated CO_2 on crops. The key messages can be summarized as follows:

First, it is very likely that climate change is likely to increase the number of people at risk of hunger compared with reference scenarios with no climate change; the exact impacts will, however, strongly depend on the projected socio-economic developments. For instance, it is estimated that climate change would increase the number of undernourished in 2080

Defining Food Security

by 5–26%, compared with no climate change or by between 5 million and 10 million (B1 SRES) and 120 million to 170 million people (A2 SRES), with within-SRES ranges depending on GCM climate projections. Using only one GCM scenario, others projected small reductions by 2080, *i.e.*, – 5%, or –10 million (B1) to –30 million (A2) people, and slight increases of +13%–26%, or H–10 million (B2) to 30 million (A1) people.

Second, it is likely that the magnitude of these climate impacts will be small compared with the impact of socio-economic development. The limitations of socio-economic forecasts, these studies suggest that robust economic growth and a decline in population growth projected for the 21st century will, in all but one scenario (SRES A2), significantly reduce the number of people at risk of hunger in 2080. At any rate, the prevalence of undernourishment will decline as all scenarios assume that world population will continue to grow to 2080, albeit at lower rates. Compared with FAO estimates of 820 million undernourished in developing countries today, several studies estimate reductions of >75% by 2080, or H–560 million to 700 million people, projecting 100 million to 240 million undernourished by 2080 (A1, B1, and B2). As mentioned, the only exception is scenario A2, where the number of the hungry is forecast to decrease only slightly by 2080; but the higher population growth rates in A2 compared with other scenarios mean that the prevalence of undernourishment will decline drastically. However, these analyses also confirm that the progress will be unevenly distributed over the developing world, and more importantly progress will be slow during the first decades of the outlook. With or without climate change, the millennium development goal of halving the prevalence of hunger by 2015 is unlikely to be realized before 2020–2030.

In addition to socio-economic pressures considered within the IPCC SRES scenarios, food production may increasingly compete with bio-energy in coming decades; studies addressing possible consequences for world food supply have only started to surface, providing both positive and negative views. Importantly, none of the major world food models discussed herein have yet considered such competition.

Third, sub-Saharan Africa is likely to surpass Asia as the most food-insecure region. However, this is largely independent of climate change and is mostly the result of the socio-economic development paths assumed for the different developing regions in the SRES scenarios. Throughout most SRES and climate-change scenarios sub-Saharan Africa accounts for 40–50% of global hunger by 2080, compared with H–24% today; in some simulations sub-Saharan Africa even accounts for 70–75% of global undernourishment by 2080. These high estimates have emerged from slower growth variants of the A2 and B2 scenarios; also an A2 variant with slower population growth yields a sharper concentration of hunger in sub-Saharan Africa. For regions other than sub-Saharan Africa, results largely depend on GCM scenarios and therefore are highly uncertain.

Fourth, although there is significant uncertainty on the effects of elevated CO_2 on crop yields, this uncertainty is carried to a much lesser extent on food security.

This result emerges from a comparison of climate change simulations with and without CO_2 fertilization effects on crop yields. As can be seen, higher CO_2 fertilization would not greatly affect global projections of hunger. In view of the fact that essentially all SRES worlds are characterized by much higher real incomes, much improved transportation and communication options, and still sufficient global food production, the somewhat smaller supplies will not be able to make a dent in global food security outcomes.

Many studies find that climate change without CO_2 fertilization would reduce the number of undernourished people by 2080 only by some 20 million to 140 million (120 million to 380 million for SRES A1, B1, and B2 without, compared with 100 million to 240 million with CO_2 fertilization effect). The exception again in these studies is SRES A2, under which the assumption of no CO_2 fertilization results in a projected range of 950 million to 1,300 million people undernourished in 2080, compared with 740 million to 850 million projected under climate change, but with CO_2 effects on crops.

Finally, recent research suggests large positive effects of climate stabilization for the agricultural sector. However, as the stabilizing effects of mitigation measures can take several decades to be realized from the moment of implementation, the benefits for crop production may be realized only in the second half of this century. Importantly, even in the presence of robust global long-term benefits, regional and temporal patterns of winners and losers that can be projected with current tools are highly uncertain and depend critically on the underlying GCM projections.

Uncertainties and Limitations

The finding that socio-economic development paths have an important bearing on future food security and that they are likely to top the effects of climate change should not, or at least not only, be interpreted as a probability-based forecast. This is because SRES scenarios offer a range of possible outcomes "without any sense of likelihood". Yet SRES scenarios, like all scenarios, do not overcome the inability to accurately project future changes in economic activity, emissions, and climate.

Second, the existing global assessments of climate change and food security have only been able to focus on the impacts on food availability and access to food, without quantification of the likely important climate change effects on food safety and vulnerability (stability).

This means that these assessments neither include potential problems arising from additional impacts due to extreme events such as drought and floods nor do they quantify the potential impacts of changes in the prevalence of food-borne diseases (positive as well as negative) or the interaction of nutrition and health effects through changes in the proliferation of vector-borne diseases such as malaria. On the food availability side, they also exclude the impacts of a possible sea-level rise for agricultural production or those that are associated with possible reductions of marine or freshwater fish production.

Third, it is important to note that even in terms of food availability, all current assessments of world food supply have focused only on the impacts of mean climate change, *i.e.*, they have not considered the possibility of significant shifts in the frequency of extreme events on regional production potential, nor have they considered scenarios of abrupt climate or socio-economic change; any of these scenario variants is likely to increase the already negative projected impacts of climate change on world food supply. Models that take into account the specific biophysical, technological, and market responses necessary to simulate realistic adaptation to such events are not yet available.

Fourth, this review finds that recent global assessments of climate change and food security rest essentially on a single modeling framework, the IIASA system, which combines the FAO/IIASA AEZ model with various GCM models and the IIASA BLS system, or on close variants of the IIASA system. This has important implications for uncertainty, given that the robustness of all these assessments strongly depends on the performance of the underlying models. There is, therefore, a clear need for continued and enhanced validation efforts of both the agro-climatological and food trade tools developed at IIASA and widely used in the literature.

Finally, we note that assessments that do not only provide scenarios, but also attach probabilities for particular outcomes to come true could provide an important element for improved or at least better-informed policy decisions. A number of possibilities are offered to address the related modeling challenges. One option would be to produce such estimates with probability-based estimates of the (key) model parameters. Alternatively, the various scenarios could be constructed so that they reflect expert judgements on a particular issue. It would be desirable to attach probabilities to existing scenarios. Information on how likely the suggested outcomes are would contribute greatly to their usefulness for policy makers and help justify (or otherwise) policy measures to adapt to or mitigate the impacts of climate change on food security.

Climate change will affect all four dimensions of food security, namely food availability (*i.e.*, production and trade), access to food, stability of food supplies, and food utilization. The importance of the various dimensions and the overall impact of climate change on food security will differ across regions and over time and, most importantly, will depend on the overall socio-economic status that a country has accomplished as the effects of climate change set in.

Essentially all quantitative assessments show that climate change will adversely affect food security. Climate change will increase the dependency of developing countries on imports and accentuate existing focus of food insecurity on sub-Saharan Africa and to a lesser extent on South Asia. Within the developing world, the adverse impacts of climate change will fall disproportionately on the poor. Many quantitative assessments also show that the socio-economic environment in which climate change is likely to evolve is more important than the impacts that can be expected from the biophysical changes of climate change.

Less is known about the role of climate change for food stability and utilization, at least in quantitative terms. However, it is likely that differences in socio-economic development paths will also be the crucial determinant for food utilization in the long run and that they will be decisive for the ability to cope with problems of food instability, be they climate-related or caused by other factors.

Finally, all quantitative assessments we reviewed show that the first decades of the 21st century are expected to see low impacts of climate change, but also lower overall incomes and still a higher dependence on agriculture. During these first decades, the biophysical changes as such

Defining Food Security

will be less pronounced but climate change will affect those particularly adversely that are still more dependent on agriculture and have lower overall incomes to cope with the impacts of climate change.

By contrast, the second half of the century is expected to bring more severe biophysical impacts but also a greater ability to cope with them. The underlying assumption is that the general transition in the income formation away from agriculture toward nonagriculture will be successful.

How strong the impacts of climate change will be felt over all decades will crucially depend on the future policy environment for the poor. Freer trade can help to improve access to international supplies; investments in transportation and communication infrastructure will help provide secure and timely local deliveries; irrigation, a promotion of sustainable agricultural practices, and continued technological progress can play a crucial role in providing steady local and international supplies under climate change.

RISKS TO FOOD SECURITY

Fossil Fuel Dependence

While agricultural output increased as a result of the Green Revolution, the energy input into the process (that is, the energy that must be expended to produce a crop) has also increased at a greater rate, so that the ratio of crops produced to energy input has decreased over time. Green Revolution techniques also heavily rely on chemical fertilizers, pesticides and herbicides, some of which must be developed from fossil fuels, making agriculture increasingly reliant on petroleum products.

Between 1950 and 1984, as the Green Revolution transformed agriculture around the globe, world grain production increased by 250%. The energy for the Green Revolution was provided by fossil fuels in the form of fertilizers (natural gas), pesticides (oil), and hydrocarbon fuelled irrigation.

David Pimentel, professor of ecology and agriculture at Cornell University, and Mario Giampietro, senior researcher at the National Research Institute on Food and Nutrition (INRAN), place in their study *Food, Land, Population and the U.S. Economy* the maximum U.S. population for a sustainable economy at 200 million. To achieve a sustainable economy and avert disaster, the United States must reduce its population by at least one-third, and world population will have to be reduced by two-thirds, says the study.

The authors of this study believe that the mentioned agricultural crisis will only begin to impact us after 2020, and will not become critical until 2050.

The oncoming peaking of global oil production (and subsequent decline of production), along with the peak of North American natural gas production will very likely precipitate this agricultural crisis much sooner than expected. Geologist Dale Allen Pfeiffer claims that coming decades could see spiraling food prices without relief and massive starvation on a global level such as never experienced before.

However, one should take note that, (numbers taken from the CIA World Factbook), the country of Bangladesh achieved food self-sufficiency in 2002 with both a far higher population density than the USA (~1000 inhabitants per square kilometer in comparison to just 30/km for the USA- so this is more than 30 times as many), and at only a tiny fraction of the USA's usage of oil, gas, and electricity.

Also, pre-industrial Chinese mini-farmers/gardeners developed techniques to feed a population of more than 1000 people per square kilometer. Hence, the dominant problem is not energy availability but the need to stop and revert soil degradation. Hybridization, genetic engineering and loss of biodiversity

In agriculture and animal husbandry, the Green Revolution popularized the use of conventional hybridization to increase yield by creating "high-yielding varieties". Often the handful of hybridized breeds originated in developed countries and were further hybridized with local varieties in the rest of the developing world to create high yield strains resistant to local climate and diseases. Local governments and industry have been pushing hybridization which has resulted in several of the indigenous breeds becoming extinct or threatened. Disuse because of unprofitability and uncontrolled intentional and unintentional cross-pollination and crossbreeding (genetic pollution), formerly huge gene pools of various wild and indigenous breeds have collapsed causing widespread genetic erosion and genetic pollution. This has resulted in loss of genetic diversity and biodiversity as a whole.

A genetically modified organism (GMO) is an organism whose genetic material has been altered using the genetic engineering techniques generally known as recombinant DNA technology. Genetically Modified (GM) crops today have become a common source for genetic pollution, not only of wild varieties but also of other domesticated varieties derived from relatively natural hybridization.

Defining Food Security

Genetic erosion coupled with genetic pollution may be destroying unique genotypes, thereby creating a hidden crisis which could result in a severe threat to our food security. Diverse genetic material could cease to exist which would impact our ability to further hybridize food crops and livestock against more resistant diseases and climatic changes.

Genetic Erosion in Agricultural and Livestock Biodiversity

Genetic erosion in agricultural and livestock biodiversity is the loss of genetic diversity, including the loss of individual genes, and the loss of particular combinants of genes (or gene complexes) such as those manifested in locally adapted landraces of domesticated animals or plants adapted to the natural environment in which they originated. The term genetic erosion is sometimes used in a narrow sense, such as for the loss of alleles or genes, as well as more broadly, referring to the loss of varieties or even species.

The major driving forces behind genetic erosion in crops are: variety replacement, land clearing, overexploitation of species, population pressure, environmental degradation, overgrazing, policy and changing agricultural systems. The main factor, however, is the replacement of local varieties of domestic plants and animals by high yielding or exotic varieties or species.

A large number of varieties can also often be dramatically reduced when commercial varieties (including GMOs) are introduced into traditional farming systems. Many researchers believe that the main problem related to agro-ecosystem management is the general tendency towards genetic and ecological uniformity imposed by the development of modern agriculture.

Intellectual Property Rights

There is much debate on whether IPRs hurt or harm independent development in terms or agriculture and food production.

Hartmut Meyer and Annette von Lossau describe both sides of the issue, while saying "Among scholars, the thesis that the impetus to self-determined development and the protection of intellectual property go hand in hand is disputed-to put it mildly.

Many studies have concluded that there is virtually no positive correlation between establishing self-sustained economic growth and ensuring protection of intellectual property rights.

Price Setting

On April 30, 2008 Thailand announces the project of the creation of the Organisation of Rice Exporting Countries with the potential to develop into a price-fixing cartel for rice.

Treating food the same as other internationally traded commodities.

On October 23, 2008, Associated Press reported the following: "Former President Clinton told a U.N. gathering Thursday [Oct 16, 2008] that the global food crisis shows "we all blew it, including me," by treating food crops "like colour TVs" instead of as a vital commodity for the world's poor....Clinton criticized decades of policymaking by the World Bank, the International Monetary Fund and others, encouraged by the U.S., that pressured Africans in particular into dropping government subsidies for fertilizer, improved seed and other farm inputs as a requirement to get aid. Africa's food self-sufficiency declined and food imports rose. Now skyrocketing prices in the international grain trade—on average more than doubling between 2006 and early 2008—have pushed many in poor countries deeper into poverty."

ESSENTIALLY OF FOOD SECURITY

At the national level, food security essentially refers to the capacity of a country to provide sufficient food for its population, and it will depend on such factors as food production, imports, food aid and within-country food distribution (in India's case, this is state-level food security). The ability to absorb shocks to such a system from drought, floods, civil strife, etc. is also relevant. At the household level, food security is here defined as access to food, that is adequate in terms of quality, quantity, safety and cultural acceptability, for all household members (Gillespie and Mason 1991). While many definitions of household security exist, they virtually all hinge on a household's ability to get sufficient food to the household door. The two main factors determining household food security - poverty and food prices - are investigated here. The cost of achieving and maintaining food security is important and can be revealed through an analysis of the proportions of either income or expenditure that is allocated to food.

Food security refers to the availability of food and one's access to it. A household is considered food-secure when its occupants do not live in hunger or fear of starvation. According to the World Resources Institute, global per capita food production has been increasing substantially for

Defining Food Security

the past several decades. In 2006, MSNBC reported that globally, the number of people who are overweight has surpassed the number who are undernourished-the world had more than one billion people who were overweight, and an estimated 800 million who were undernourished. According to a 2004 article from the BBC, China, the world's most populous country, is suffering from an obesity epidemic. In India, the second-most populous country in the world, 30 million people have been added to the ranks of the hungry since the mid-1990s and 46% of children are underweight.

Worldwide around 852 million people are chronically hungry due to extreme poverty, while up to 2 billion people lack food security intermittently due to varying degrees of poverty. Six million children die of hunger every year-17,000 every day. As of late 2007, export restrictions and panic buying, US Dollar Depreciation, increased farming for use in biofuels, world oil prices at more than $100 a barrel, global population growth, climate change, loss of agricultural land to residential and industrial development, and growing consumer demand in China and India are claimed to have pushed up the price of grain. However, the role of some of these factors is under debate. Some argue the role of biofuel has been overplayed as grain prices have come down to the levels of 2006. Nonetheless, food riots have recently taken place in many countries across the world.

It is becoming increasingly difficult to maintain food security in a world beset by a confluence of "peak" phenomena, namely peak oil, peak water, peak phosphorus, peak grain and peak fish. Approximately 3.3 billion people, more than half of the planet's population, live in urban areas as of November 2007. Any disruption to farm supplies may precipitate a uniquely urban food crisis in a relatively short time. The ongoing global credit crisis has affected farm credits, despite a boom in commodity prices. Food security is a complex topic, standing at the intersection of many disciplines.

A new peer-reviewed journal of *Food Security: The Science, Sociology and Economics of Food Production and Access to Food* began publishing in 2009. In developing countries, often 70% or more of the population lives in rural areas. In that context, agricultural development among smallholder farmers and landless people provides a livelihood for people allowing them the opportunity to stay in their communities. In many areas of the world, land ownership is not available, thus, people who want or need to farm to make a living have little incentive to improve the land.

In the US, there are approximately 2,000,000 farmers, less than 1% of the population. A direct relationship exists between food consumption levels and poverty. Families with the financial resources to escape extreme poverty rarely suffer from chronic hunger, while poor families not only suffer the most from chronic hunger, but are also the segment of the population most at risk during food shortages and famines.

Two commonly used definitions of food security come from the UN's Food and Agriculture Organization (FAO) and the United States Department of Agriculture (USDA):

- Food security exists when all people, at all times, have physical, social and economic access to sufficient, safe and nutritious food to meet their dietary needs and food preferences for an active and healthy life.
- Food security for a household means access by all members at all times to enough food for an active, healthy life. Food security includes at a minimum (1) the ready availability of nutritionally adequate and safe foods, and (2) an assured ability to acquire acceptable foods in socially acceptable ways (that is, without resorting to emergency food supplies, scavenging, stealing, or other coping strategies). (USDA)

The stages of food insecurity range from food secure situations to full-scale famine. "Famine and hunger are both rooted in food insecurity. Food insecurity can be categorized as either chronic or transitory. Chronic food insecurity translates into a high degree of vulnerability to famine and hunger; ensuring food security presupposes elimination of that vulnerability. [Chronic] hunger is not famine. It is similar to undernourishment and is related to poverty, existing mainly in poor countries."

ENSURING GLOBAL FOOD SECURITY

Eradicating malnutrition across the globe will require an ambitious and coherent mix of policies, covering agriculture, trade, aid, environmental protection, research, investment and more. The correct balance will help provide the conditions for a general rise in prosperity — which is what puts food in people's mouths.

At the World Food Conference in Rome in 1974, the world's governments examined global food production and consumption, concluding their meeting with the solemn declaration that 'every man,

Defining Food Security

woman and child has the inalienable right to be free from hunger and malnutrition in order to develop their physical and mental faculties'.

In spite of the considerable progress in increasing food production and reducing malnutrition over the two intervening decades, those ambitious words have yet to become a reality: the Food and Agriculture Organisation of the United Nations (FAO) estimates that over 800 million people in the developing world alone are still undernourished. Indeed, the FAO fears that, unless action is taken, many of the current problems of food security will persist and some will become worse. In Africa in particular, lagging food production and a rapidly growing population are increasing the risk of serious and persistent food shortages.

At the World Food Summit held, again in Rome, from 13 to 17 November 1996, the member governments of the FAO reaffirmed their commitment to the eradication of hunger and agreed on a plan of action for increasing the year-round availability of food, improving access to supplies, and reducing the number of malnourished people.

To guarantee that all individuals have access to sufficient food — their 'food security' — the world must produce enough to feed its expanding population; and that output must be readily available to all. It is therefore literally vital that government policies do not impair incentives to produce and market food. Many poorer countries, in attempting to keep costs low for urban consumers, depress the prices received by farmers and actually worsen the supply of food.

Developing Food Resources

With enough investment to support the development and application of improved methods of production, the world can indeed produce the quantities required, even though its population is projected to rise from roughly 5.8 billion currently to 7.6 billion by the year 2020. In recent decades world production of food has expanded at a rapid rate, aided from the 1960s onwards by a scientific revolution that has led to major gains in the yields of rice and wheat.

The introduction of improved varieties of crops and the use of fertilizer, pesticides, mechanisation and irrigation have resulted in production increases in excess of population growth in many regions of the world. Increases in yields per hectare of 3% or more per year have not been uncommon. Science and technology have been of paramount importance, first, by making possible the dietary improvements that have accompanied the expansion of supply, and, more generally, in promoting economic development by lowering the real cost of food and releasing labour from agriculture for employment in other sectors.

In many richer countries, the speed with which agriculture is expected to adjust to changes in productivity has been a contentious political issue. Attempts have been made to protect farm incomes by supporting prices and even to limit necessary structural changes, such as an increase in farm size. Many of these attempts have been costly failures. They may even have accelerated the very pressures for structural change that they were intended to control.

As supply continues to expand, more attention will have to be given to the protection of the environment, which is now coming under considerable pressure from intensive farming techniques. Most of the world's available cropland is currently being used, and little further production can be expected from bringing additional land under cultivation. The increases in crop yields necessary to meet demand will have to be both economically and environmentally sustainable. Further advances in biotechnology, such as the genetic modification of plants to reinforce their resistance to pests or diseases, offer

Substantial promise for increasing yields with minimal impact on the environment. Improvements in the management of scarce natural resources, particularly water, are important. Irrigated agriculture contributes nearly 40% of world food production from only 17% of the world's cultivated land, but accounts for 70% of global water use.

Nor are these concerns purely terrestrial. Fish are an important source of protein, particularly in developing countries, with a vital role to play in food security. But stocks are under considerable pressure because of over-fishing. A plan of action for the sustainable exploitation of fisheries was agreed in Kyoto, Japan, in December 1995, by 95 member governments of the FAO.

It calls for improved international co-operation and co-ordination in all aspects of fisheries management, not least better monitoring of stocks, strengthening the scientific basis for their management, and the use of fishery conservation and management arrangements to reduce over-fishing. If such a route is not followed, there is a risk of a large deficit in supply early in the next century. That would hit the developing countries particularly hard.

The expansion of supplies through aquaculture could be important

Defining Food Security

in filling the gap. The FAO estimates that world aquaculture production is likely to more than double by the year 2010, and could then account for over 35% of fish consumed. But so far aquaculture has centred largely on high-value species, such as salmon and shrimp, which are unlikely to provide inexpensive protein for developing countries. Their best option to increase the supply of protein is to focus on the promotion of aquaculture as part of sustainable agricultural systems by, for example, the rearing of fish in rice paddies.

The creation of sustainable agricultural systems is a complex process that requires appropriate technology and policies, and attention to education and training. The effective diffusion of improved techniques is crucial in increasing output and minimising post-harvest losses. And since many of the world's food-producers are women, equal access for both men and women to training in these techniques is extremely important.

The returns from agricultural research are high, yet investment in research, particularly in developing countries, has been stagnant in the 1990s. That calls for a higher priority for funding, both nationally and internationally. Closer co-operation between the private and public sector — through, for example, joint ventures and the protection of intellectual property — would help to ensure that sufficient resources are allocated to research in the future. The OECD countries can help here: their extensive R&D infrastructure and scientific expertise can be used to strengthen agricultural research programmes in poorer countries, particularly through co-operative projects and training.

Economic Growth V. Poverty

Food insecurity can be a vicious circle: people who are malnourished' often do not have the resources to grow their own food, or enough income to purchase it from others. The solution clearly lies in increasing the incomes of the poorest. In some countries that means ensuring that poor people have access to sufficient firmland and capital. In others it means finding ways to promote the growth of economic activity in rural and urban areas. The creation of a policy environment that encourages private investment and capital-formation throughout the economy is an important ingredient in accelerating economic growth and the creation of employment. All governments have to direct their efforts towards ensuring the necessary conditions for sustainable economic growth, which brings with it an improved distribution of income.

Political instability often contributes to food insecurity. Participatory and pluralistic political systems, with governments responsive and responsible to their peoples, are thus most conducive to food security. The disruption of food supplies through wars and civil strife has been a major factor in causing malnutrition and famine.

By contrast, economic growth and political stability in many parts of the world have done much to improve food security. It is not coincidental that where new techniques have made the largest contribution to agricultural productivity, in many of the large rice-producing countries of Asia, is also where the fastest economic growth has been occurring. Other regions of the world, particularly Africa, which have not had such a favourable combination of suitable technologies and political stability, have seen their food supplies deteriorate under the pressure of rapid population growth. It is towards such areas in particular that the focus of improving food security must be directed.

11

Meat and Poultry Hazard Analysis

The Hazard Analysis Critical Control Point approach is a system of preventative process controls that is widely recognized by scientific authorities nationally and internationally and used throughout the food industry to produce products in compliance with health and safety requirements. About five years ago, the U.S. Department of Agriculture decided to apply HACCP to meat and poultry products as a result of a major policy change that focused USDA efforts more effectively on risks associated with pathogens.

The implementation of the system is expected to be complete in January 2000. Since the establishment of this rule, the incidence of food-borne illness associated with meat and poultry products continues to be monitored and impacts thus far appear favourable.

The need to focus more heavily on pathogenic microorganisms, and to implement preventive approaches such as HACCP, was established and supported by studies conducted over the past 15 years by the National Academy of Sciences, the Government Accounting Office, and the USDA. In 1994, the Council for Agricultural Science and Technology estimated that 6.5 to 33 million cases of food-borne illness and up to 9,000 deaths occur each year because of food-borne illness and related problems. However, public support for change in the food safety system did not truly begin to emerge until the 1993 outbreak of food-borne illness associated with *Escherichia coli* O157:H7 in undercooked hamburgers. Thus, a comprehensive strategy for change was developed with HACCP and pathogen reduction as the centerpiece.

The pathogen reduction and HACCP rule consists of four mandatory provisions. First, it requires all plants to have standard operating procedures for sanitation. The second provision requires slaughter plants to test carcasses for generic *Escherichia coli*, an indicator of fecal contamination. Third, all meat and poultry plants must implement HACCP systems as a means of preventing or controlling contamination from pathogens, as well as other hazards. Under HACCP, plants identify and evaluate the hazards that could affect the safety of their products and institute controls necessary to prevent those hazards from occurring or at a minimum, keep them within the acceptable limits. Finally, to make sure HACCP systems are working as intended, the rule mandates performance standards for salmonella at slaughter and grinding plants.

One of the key guiding principles of the rule was the clarification that industry is responsible for producing and marketing products that are safe, unadulterated, and properly packaged and labelled. With industry assuming its proper responsibility, federal agencies can use limited resources more efficiently and effectively. Another key concept was the combination of HACCP and performance standards for pathogen reduction. HACCP has to be combined with objective means of verifying food safety compliance. Up until this time, microbial performance standards for raw products, with the exception of *Escherichia coli* O157:H7 in ground beef, had not been established. The development of these rules was broadly supported by the public through numerous public meetings soliciting input.

Although the implementation of HACCP is still underway, preliminary data indicates positive results. Salmonella prevalence in broilers, swine, ground beef, and ground turkey was significantly lower after HACCP implementation than in the baseline studies that were conducted before implementation. Data released this year from the FoodNet Active Surveillance System for food-borne illness show that during 1998, the rate of campylobacter, salmonella, and cryptosporidium infections declined nationwide, and that *Salmonella enteritidis* infections declined in all states but Oregon.

Concurrent with HACCP and its implementation within meat and poultry slaughter and processing establishments, USDA's broad food safety strategy addresses every step in the food production process, from animals on the farm, to slaughter and processing, and to product distribution and preparation. The USDA is working through the Partnership for Food Safety and Education and the Fight BAC Campaign to ensure that consumers know how to properly prepare, handle, and store foods. Additionally, USDA is implementing many regulatory reform initiatives to convert traditional regulations to performance standards, thus clarifying the roles and responsibilities of industry and allowing them more flexibility to develop and introduce new technologies to improve food safety.

Many lessons have been learned throughout this process. First is the need to base changes on the best science available and then make adjustments as new information becomes available. Second is the importance of making HACCP mandatory for all plants to ensure that consumers receive safe food regardless of the size of the establishment from which it originates. The third lesson learned is the need to implement HACCP in the context of a broad farm-to-table strategy to enable the far-reaching changes that are necessary to effectively reduce the incidence of food-borne illness. Finally, it is important to have public participation in the process to ensure the discovery and application of the best possible scientific solutions available.

HACCP, as part of a broad-based food safety strategy, is working as intended to reduce the incidence of food-borne illness associated with meat and poultry products. Although many scientific gaps continue to exist that prevent the establishment of "pure" public health standards, revisions in the standards will occur, as new data becomes available. For the future, USDA plans to improve efforts to quantitate the public health risks associated with certain pathogens and foods and ensure that ongoing research will provide sufficient and accurate information on which to base regulatory decisions and develop new preventative procedures. To the extent possible, regulators must continue to focus efforts on designing policies and focusing resources on the most immediate and significant public health risks.

Escherichia Coli O157:H7 and Listeria Monocytogenes

Over the past two decades, the Centres for Disease Control and Prevention (CDC), the Food and Drug Administration (FDA), and the U.S. Department of Agriculture (USDA) have been conducting investigative studies to determine the origin and methods of prevention for two food-borne pathogens, *Escherichia coli* O157:H7 and *Listeria monocytogenes*. Many technologies have been developed, preventive processes implemented, and lessons learned throughout this process. Following are case studies on these pathogens, with brief overviews of what the agencies discovered, how they responded to these findings, and ideas about what should be done in the future to prevent and reduce the incidence of these and other food-borne illnesses.

Escherichia coli O157:H7

The human pathogen *Escherichia coli* O157:H7 emerged as threat in 1982 when it was identified by CDC in two outbreaks associated with ground beef sandwiches. In a series of similar outbreaks over the next 10 years, CDC was able to determine several characteristics about the organism, most importantly that cattle were a principal source or host carrier and that contaminated cattle manure was likely the source of many infections. Studies also revealed that *Escherichia coli* O157:H7 caused hemorrhagic colitis, had no unusual heat tolerance and could be controlled by proper cooking temperatures, had a very unusual acid tolerance, and could survive fermentation of meat.

However, it was not until 1993, after a serious outbreak involving more than 700 illnesses and four deaths from eating undercooked hamburgers, that FDA changed the Food Code, which previously recommended cooking temperatures for ground beef that were insufficient to kill large populations of *Escherichia coli* O157:H7. The new Food Code criteria indicated that ground beef patties be cooked to an internal temperature of 155 degrees for 15 seconds. In addition, USDA required that safe handling labels be used for raw meat and poultry products. A year later, USDA declared *Escherichia coli* O157:H7 an adulterant in raw ground beef, established a zero tolerance, and initiated end product testing for raw ground beef. However, food-borne illness continued to increase and it was clear that further action was necessary. Consequently, between 1995 and 1997, USDA published a rule on pathogen reduction, FDA approved irradiation of red meat, and CDC introduced the FoodNet system, an active surveillance system that monitors the occurrence of illnesses associated with selected food-borne pathogens. Additionally, CDC, USDA, and the food industry initiated the Fight BAC campaign, a programme which educates consumers about proper food handling techniques in the home. In 1998, USDA published a document recommending that a thermometer be used to measure the temperature of cooked ground beef patties. Currently, regulatory agency efforts to reduce food-borne illness due to *Escherichia coli* O157:H7 continue, with USDA planning to implement HACCP in all meat processing plants by the year 2000.

Bibliography

Asiedu, J.J.: *Processing chicken*, MacMillan Press Ltd, London, 1989.
Axtell, B., Kocken E., and Sandhu, R., *Fruit Processing*, IT Publications, London, 1993.
Barlow K: *Ecology of Food and Nutrition*, Oceania, New Guinea, 1984.
Bindon JR: *Food and Foodways*, Oceania, Samoa, 1988.
Cammack, J.: *Basic Accounting for Small Groups*, Oxfam, Oxford, 1992.
Casado, Matt A: *Food and Beverage Service Manual*. New York: Wiley, 1994.
Coultate, T.P.: *Food: the Chemistry of its Components*, Burlington House, London, 1984.
Fellows, P.J.: *Food Processing Technology*, Woodhead Publishing, Cambridge, 1993.
Ferrando, R.: *Traditional and Non-Traditional Foods*, FAO Publications, Rome, Italy, 1981.
Garg, G.P. : *Microbial Biochemistry*, Discovery Pub, Delhi, 2010.
Gaskell G: *Biotechnology-the Making of a Global Controversy*, Cambridge, Cambridge University Press, 2002.
George Kovics: *Progress in Mycology*, Scientific, Delhi, 2010.
Gour, H.N. : *Integrated Plant Pathology*, Scientific, Delhi, 2009.
Gowen S.: *Banana and Plantains*, London, Chapman Hall, 1996.
Gupta, Preeti : *Immunology and Microbiology*, Pointer Pub, Delhi, 2008.
Harper, M: *Small Business in the Third World*, J Wiley & Sons Ltd, Chichester, 1984.
Hays, Judi Radice: *Restaurant & Food Graphics*. Glen Cove: PBC International, 1994.
Herminie Broedel Kitchen: *Soils and Crops : Diagnostic Techniques*, Satish Serial Publishing, Allahabad, 2004.
Hidellage, V.: *Making Safe Food*, IT Publications, London, 1992.
Hobbs, B. and Roberts, D.: *Food Poisoning and Food Hygiene*, Edward Arnold Ltd, London, 1987.
Howard, J.: *Safe Drinking Water*, Oxfam Technical Guide, Oxfam, 1979.
Jagtiani, J., Chan, H.T. and Sakai, W.S.: *Tropical Fruit Processing*, Academic Press, San Diego, California, 1988.
Jasra, O P : *Techniques in Microbiology*, Sarup, Delhi, 2004.
Jay, J.M.: *Modern Food Microbiology*, D van Nostrand, New York, 1978.
Jha, D.K. : *Laboratory Manual on Plant Pathology*, Pointer, Delhi, 2004.

Jones, R. M.: *Plant Resources of South-East Asia,* Wageningen, Pudoc Scientific Publishers, 1992.
Kannaiyan, S. : *Rice Management Biotechnology,* Associated, Delhi, 1995.
Kanthaliya, P C : *Soils and Fertilisers at a Glance,* Agrotech, Delhi, 2004.
Kataria, T N : *Plant and Crop Physiology,* Pearl Books, Delhi, 2008.
Keshav Prasad Yadav: *Application of Morphometry in Geomorphology,* Radha Pub, Delhi, 2008.
Khan, Samiullah: *Plant Breeding Advances and in vitro Culture,* CBS, Delhi, 1997.
Khatri, N.K. and Manish Pathak: *Fundamentals of Plant Pathology,* Agrobios, Delhi, 2006.
Kumar, Arvind : *Environmental Pollution and Agriculture,* APH, Delhi, 2002.
Kumar, Vinod : *Biotechnology and Plant Pathology : Current Trends,* Oxford Book Company, Delhi, 2009.
Kurzweil, Ray: *The Age of Spiritual Machines,* New York, Penguin Books, 1999.
Lal. Madan : *Physiology, Biochemistry and Biotechnology,* Manglam Pub, Delhi, 2007.
Lance, J.W. : *Migraine and Other Headaches,* New York, U.S.A: Scribner, 1986.
Lata Bhattacharya: *Animal Biochemistry,* Discovery, Delhi, 2010.
Leena Parihar: *Advances in Applied Microbiology,* Agrobios, Delhi, 2008.
Low, J.: *Fruit and Vegetables,* IT Publications, London, 1984.
Lyster, S. : *Animals and Their Moral Standing.* London : Routledge, 1997.
MacDonald, I. and Low, J.: *Fruit and Vegetables,* IT Publications, London, 1984.
Madan Lal Bagdi: *Physiology, Biochemistry and Biotechnology,* Manglam Pub, Delhi, 2007.
Mathialagan, P : *Textbook of Animal Husbandry and Livestock Extension,* International Book Distributing Co, Delhi, 2005.
Metting, Jr., F.B.: *Soil Microbial Ecology: Applications in Agricultural and Environmental Management,* Marcel Dekker, New York, 1992.
Miller, Henry I. : *The Frankenfood Myth: How Protest and Politics Threaten the Biotech Revolution,* New York, Praeger, 2004.
Mishra, Akhilesh : *Plant Pathology : Diseases and Management,* Agrobios, Delhi, 2005.
Mondal, M.S. : *Pollen Morphology and Systematic Relationship of the Family Polygonaceae,* BSI, Delhi, 1997.
Mondal, Sumit: *Pollen Morphology and Systematic Relationship of the Family Polygonaceae,* BSI, Delhi, 1997.
Montgomery, G. G.: *The Early Placental Mammal Radiation Using Bayesian Phylogenetics,* Science, December 2001.
Murray, David: *Seeds of Concern: The Genetic Manipulation of Plants,* Sydney, University of New South Wales, 2003.
Murty, S. : *Faces and Phases of Agriculture and Industry in India,* RBSA Pub, Delhi, 2004.
Murugesan, R. : *Energy Use Efficiency in Dryland Agriculture,* Kalpaz, Delhi, 2010.
Nair, L.N. : *Vistas in Mycology and Plant Pathology,* Commonwealth, Delhi, 2000.

Bibliography

Narvekar, Raghunath : *Molecular Biochemistry : Principles and Practices,* Adhyayan Pub, Delhi, 2008.

Ngoddy, P.O., : *Integrated Food Science and Technology for the Tropics,* Macmillan Press Ltd., London, 1985.

Nobel, P. S.: *Physicochemical and Environmental Plant Physiology,* Academic Press, San Diego, 1999.

Nyholt, D.H. : *The Vitamin & Herb Guide,* Alberta, Canada: Global Health Ltd., 1992.

Pollock NJ: *The Concept of Food in Pacific Society: A Fijian Example,* Oceania, Fiji, 1985.

Rauch, G.H.: *Jam Manufacture,* Leonard Hills Books, London, 1965.

Roberts, D.: *Food Poisoning and Food Hygiene,* Edward Arnold Ltd, London, 1987.

Sims-Bell, Barbara: *Career Opportunities in the Food and Beverage Industry.* New York: Facts on File, 1994.

Sprenger, R.A.: *The Food Hygiene Handbook,* Highfield Publications, Doncaster, 1996.

Tindall, H.D.: *Vegetables in the Tropics,* MacMillan Press Ltd., London, 1983.

Tressler, D.T.: *Fruit and Vegetable Juice Processing,* AVI Publications, Conn., 1982.

Zaccarelli, Herman E.: *Food Service Management by Checklist: A Handbook of Control Techniques.* New York: Wiley, 1991.

Index

A

Agricultural Production, 255, 256, 258, 261.
Agricultural Systems, 257.
Alcohol Permanently, 141.
Allergic Reactions, 66, 67, 68, 180, 185, 199.
Antibodies, 3, 5, 44, 113, 114, 122, 131.
Autoimmune Diseases, 117.

B

Bacillus Cereus, 16, 46, 93, 213.

C

Campylobacter, 152.
Carbohydrates, 24, 41, 62, 64, 84, 87, 88, 89, 93, 95, 96, 97, 104, 110, 165.
Cattle Manure, 268.
Chemical Hazards, 198, 199, 202, 203.
Chryseobacterium, 9, 10, 11, 12, 14, 85, 86, 87, 88, 89, 90, 91, 92, 102, 103, 104, 105, 106.
Civilization Necessitates, 142.
Clostridium Botulinum, 28, 29, 44, 95, 212, 215.
Clostridium Perfringens, 17, 212, 215.
Compositional Comparison, 60.

Concentrated Milk Products, 160.
Contamination, 134.
Cryptosporidium Parvum, 215.

D

Dairy Foods, 23, 134, 157.
Dairy Products, 11, 16, 22, 85, 99, 105, 107, 108, 209, 213, 217, 218.
Degradation, 74, 231.
Disaccharides, 88, 95.

E

Enumeration of Cells, 43.
Environmental Factors, 67, 95, 99.
Environmental Safety, 190, 196.
Enzymes, 10, 85, 95, 97, 98, 99.
Essentially of Food Security, 258.
Ethnic Groups, 205.
Experimental Design, 66.

F

Fatty Acids, 64, 85, 88, 96, 97, 98, 104, 108, 165, 166, 200.
Fermentation, 29.
Fermented Foods, 24.
Fermented Milk, 23, 158, 159.
Field Verification, 80.
Fish Species, 188.
Fluid Milk Products, 156.
Food Authenticity, 213.
Food Borne Bacteria, 14, 15.

Index

Food Chain, 49, 52, 189, 202.
Food Components, 38, 56, 57, 65, 66, 67, 72, 73, 74, 76, 182, 203.
Food Handlers, 18, 22, 53, 206, 217.
Food Oils, 165, 166.
Food Preservation, 24, 25, 26, 27, 127, 221, 222.
Food Prices, 256, 259.
Food Processing, 17, 31, 35, 53, 72, 166, 229.
Food Production, 45, 53, 74, 189, 192, 198, 225, 228, 238, 239, 240, 241, 242, 243, 246, 248, 251, 252, 257, 258, 260, 261, 262, 266.
Food Safety, 7, 8, 9, 49, 55, 56, 69, 73, 74, 75, 76, 77, 78, 79, 80, 81, 82, 179, 180, 187, 190, 196, 200, 202, 203, 205, 211, 214, 221, 236, 246, 247, 253, 265, 266, 267.
Food Safety Considerations, 55.
Food Security, 190, 224, 225, 226, 227, 228, 232, 233, 234, 235, 236, 237, 238, 239, 240, 241, 243, 244, 245, 246, 247, 248, 249, 250, 252, 253, 254, 255, 257, 258, 259, 260, 261, 262, 263, 264, 266, 267, 268.
Food Spoilage, 10, 11, 25, 26, 27, 84, 85, 92, 93, 100, 107, 111.
Food Supply, 58, 59, 60, 67, 73, 127, 128, 171, 179, 185, 190, 196, 198, 199, 203, 204, 206, 235, 240, 243, 248, 251, 253.
Foodborne Disease, 49, 128, 202, 203, 204, 205, 210, 220, 221.
Foodborne Pathogens, 19, 202, 208, 214.
Fossil Fuel Dependence, 255.
Fruitarians, 143.

G

Gene Therapy, 126, 173, 176.
Gene Transfer, 56, 61, 69, 70, 71, 169, 172, 184, 185, 188, 194.
Genetically Modified Animals, 73, 187.
Genetically Modified Microorganisms, 71, 72, 214.
Genetically Modified Organism, 57, 58, 59, 60, 61, 69, 71, 74, 75, 171, 173, 256.
Genetically Modified Plants, 67, 69, 70, 71, 75, 176.

H

Halts Spoilage, 27.
Hazard Classification, 45.
Home Canning, 27, 29.
Human Health, 74, 115, 118, 180, 189, 190, 191, 192, 197, 202.
Human Rights, 237.
Human Sources, 50.
Hydrolytic Activities, 11, 12.
Hypersensitivity Reactions, 187, 197.

I

Immune System, 16, 66, 121, 185, 187, 197, 209, 210, 218.
Industrial Chemicals, 74, 75.
Infectious Dose, 7, 17, 19, 212.

L

Life Processes, 40, 41.
Lipids, 96.
Listeria Monocytogenes, 16, 49, 131, 211, 217, 267.

M

Microbial Deterioration, 38.
Microbial Determinations, 223.
Microbiological Control, 50.
Monoclonal Antibodies, 44.

Monocytogenes, 114, 131, 132.
Monosaccharides, 88, 95.

N

Nitrogen Sources, 41.

P

Pasteurised Milk Products, 150, 156.
Pasteurization, 96.
Pasteurized Food, 49.
Pathogenicity of Microorganisms, 72.
Pathogens, 107, 114, 131.
Pharmaceuticals, 74, 75.
Phenotypic Characteristics, 57, 59, 103.
Physical Hazards, 45, 50, 202, 203.
Plant Protein, 166.
Polysaccharides, 85, 88, 89, 95.
Presence of Enzymes, 26.
Preserve Foods, 24, 27, 221.
Protein Quality, 166, 167, 168.
Proteins, 41, 61, 62, 63, 64, 67, 68, 69, 71, 76, 85, 94, 96, 97, 99, 100, 107, 110, 112, 120, 121, 124, 158, 167, 168, 169, 173, 176, 183, 185, 186, 187, 188, 189, 197, 200, 213, 222.
Protozoan Parasites, 208.
Ptomaine Poisoning, 7.

R

Raw Food, 16, 47, 49, 50, 52, 54.
Raw Milk, 104, 105, 106, 109, 133, 134, 135, 148, 151, 152, 153, 154, 155, 156, 157, 211, 214, 216, 219.
Risks to Food Security, 255.

S

Safety Assessment, 56, 57, 58, 61, 62, 63, 64, 72, 73, 74, 76, 178, 179, 180, 181, 182, 183, 184, 185, 187, 191, 196, 197.
Sanitary Operations, 35.
Sensory Analysis of Milk, 13.
Spoilage in Milk, 133.
Spoilage of Cheeses, 157.
Staphylococcus, 113, 130.
Staphylococcus Aureus, 46, 54, 113, 118, 130, 210, 217.
Substantially Equivalent, 56, 57, 58, 61, 65, 66, 74, 75.
Sustainable Development, 241.

T

Toxoplasma Gondii, 218.
Transgenic Animals, 174, 175, 187.
Transgenic Microbes, 175, 176.
Transgenic Plants, 169, 176, 191.

V

Verotoxigenic Escherichia Coli, 152.

W

Water Bath Canner, 28.
Water Management, 247.
Wild Salmon, 193.

Y

Yersinia Enterocolitica, 153, 209.

❏❏❏